Wim J.C. Melis

Computing: The Current and its Probability Based Future

Wim J.C. Melis

Computing: The Current and its Probability Based Future

Where Current Computers struggle and how Probability Based Computing can overcome this

LAP LAMBERT Academic Publishing

Impressum / Imprint
Bibliografische Information der Deutschen Nationalbibliothek: Die Deutsche Nationalbibliothek verzeichnet diese Publikation in der Deutschen Nationalbibliografie; detaillierte bibliografische Daten sind im Internet über http://dnb.d-nb.de abrufbar.
Alle in diesem Buch genannten Marken und Produktnamen unterliegen warenzeichen-, marken- oder patentrechtlichem Schutz bzw. sind Warenzeichen oder eingetragene Warenzeichen der jeweiligen Inhaber. Die Wiedergabe von Marken, Produktnamen, Gebrauchsnamen, Handelsnamen, Warenbezeichnungen u.s.w. in diesem Werk berechtigt auch ohne besondere Kennzeichnung nicht zu der Annahme, dass solche Namen im Sinne der Warenzeichen- und Markenschutzgesetzgebung als frei zu betrachten wären und daher von jedermann benutzt werden dürften.

Bibliographic information published by the Deutsche Nationalbibliothek: The Deutsche Nationalbibliothek lists this publication in the Deutsche Nationalbibliografie; detailed bibliographic data are available in the Internet at http://dnb.d-nb.de.
Any brand names and product names mentioned in this book are subject to trademark, brand or patent protection and are trademarks or registered trademarks of their respective holders. The use of brand names, product names, common names, trade names, product descriptions etc. even without a particular marking in this work is in no way to be construed to mean that such names may be regarded as unrestricted in respect of trademark and brand protection legislation and could thus be used by anyone.

Coverbild / Cover image: www.ingimage.com

Verlag / Publisher:
LAP LAMBERT Academic Publishing
ist ein Imprint der / is a trademark of
OmniScriptum GmbH & Co. KG
Heinrich-Böcking-Str. 6-8, 66121 Saarbrücken, Deutschland / Germany
Email: info@lap-publishing.com

Herstellung: siehe letzte Seite /
Printed at: see last page
ISBN: 978-3-659-67356-6

Copyright © 2015 OmniScriptum GmbH & Co. KG
Alle Rechte vorbehalten. / All rights reserved. Saarbrücken 2015

Dedicated to all those who inspire and support me.

Contents

1 Introduction 7

2 Mainstream Computing 9
   2.1 "Computing" in Current Computers 10
      2.1.1 VLSI Manufacturing 10
      2.1.2 Application Specific Integrated Circuits 18
      2.1.3 Reconfigurable Hardware 19
      2.1.4 General Purpose Processors 22
      2.1.5 Heterogeneous Computing 25
   2.2 Data-Flow within Existing Platforms 26
      2.2.1 Hardware Implementation 30
      2.2.2 Application Specific Integrated Circuit 32
      2.2.3 Reconfigurable Architectures 33
      2.2.4 General Purpose Computer 36

3 Intelligent Hardware 41
   3.1 What is "Intelligent" Computing 42
   3.2 The Need for Intelligent Computing 44
   3.3 Current Systems for Intelligent Computing 51

4 Probability Based Computing — 59
  4.1 The Applications 60
  4.2 How to Achieve Probability Based Computing 61
  4.3 How the Functionality will be Delivered 66

5 Relevant Computing Developments — 73
  5.1 Bayesian Networks 74
  5.2 Cellular Automata 76
  5.3 Quantum Computing 81
  5.4 Artificial Intelligence 84
  5.5 Neural Networks 85
  5.6 Reservoir Computing 87
  5.7 Reversible Computing 88

6 The Future — 91
  6.1 Designing with / for Imperfections 92
  6.2 Information Density 93
  6.3 Feedback 95
  6.4 Timing 96
  6.5 Evolution Algorithms 98
  6.6 How to Achieve the Change 99

7 Conclusion — 103

Bibliography — 109

List of Figures

2.1 Overview of Passive Elements 17

3.1 Overview of General Hierarchical Temporal Memory Network (a) and each of the Specific Nodes (b). 48
3.2 Patterns that would be Stored / Ignored for Storage in the Spatial Pooler 49
3.3 Hierarchical Temporal Memory 2-D (a) and 3-D (b) Network 54
3.4 Hierarchical Temporal Memory Hardware Implementation Structure for 6-Level Binary Hierarchical Structure, where each Node is Composed of a Set of 3 Nodes, each with Spatial and Temporal Pooler. 57

5.1 Cellular Automata Rule Set 78
5.2 Reservoir Computing Structure 88

List of Tables

2.1 Performance and Area for Heterogeneous Implementation (Search-Size ± 4) 37

Preface

Have you ever come across a "feature" created on this planet, which did not quite please you. Well, we all do, but the bad news is that a lot of what is around has been created by us, humans, with the goods news being that we can consequently change it again for the better. Now, if you failed to provide an answer to the first question, then maybe you overlooked administration going all the way to bureaucracy ? just to give one example.

This book is not going to solve the administration problem, although surely there are some people out there that are happy to deal with these. My background is different, and they say it is a person's history that makes him / her. My background is scientific / technical, so I shall share with you one of my "frustrating features", namely the fact that we have to "measure / express" everything with numbers. Most would claim it makes things easier to understand, which is true to some extend, but there are things that can and things that cannot be measured. Even for the latter, attempts are made to express things in numbers, with all the necessary consequences. So, if you are after numbers, equations and all the like, you should probably put this book back in the shelf to avoid disappointment, but if you are looking for a safe heaven from the "number-mad" world in which we live, then you are definitely in the right place. After all, numbers and mathematics have their limitations, just like any language, and if you ever studied another language, then you will be well aware that some expressions are very efficient in one language, maybe a few words only, but one may need several sentences to express the same in another language. Quite often the reason for this difference in efficiency lies at some point in history, but that would bring us to far from the point itself. Languages have efficiencies and inefficiencies, and so they also have their limitations. The same applies to the language of numbers and formula, namely maths. So, do we really want to, or can we express it all by using maths ? Indeed, no ! So, why try ? At the end of the day, an idea is generally first expressed in words before ever being grasped in maths / numbers, with the role of numbers generally being that they "proof" whether things are as good / bad as originally stated, no more nor less.

As an engineer, one tends to develop a certain "intuitive feeling" of what will work and what will not, and that allows for decisions to be made, even way before there are any numbers involved. So, if you think there should be numbers / formula, then imagine they are there, or create them yourself. As for the book itself, the main focus will be on making you understand what the problems are that current computing technology faces and how and why our thinking will need to change drastically to overcome these existing problems. As you will have understood by now, this book looks at the larger picture of technological challenges / changes and how and were we should be going. You will notice that looking at the larger picture allows one to move on much more quickly from one idea to another, and hopefully, by the end of the journey you have also learned to appreciate living without maths as the definitive, and for some, only, language of science and engineering.

The main aim of this book is to stimulate thinking and introduce some new concepts. It has therefore been written to target a rather large audience, since it is often easier to train someone with fresh ideas rather than retrain someone who has been thinking a certain way for a long period of time. A scientific / engineering / computer science background will help, but should not be essential. The journey itself will start by focusing on what is wrong with the current approaches, and that from this generic viewpoint. As much as you may be aware of some of these problems, it may be wise not to skip this section, as it brings you a better understanding of the thinking process that will continue throughout the book itself. Once the problems have been covered, the next section will look at what the current approaches are that people would take, based on existing technology. Following that, there will be a bit more "dreaming" of the future, and it will be up to us to ensure this does not remain "just dreaming", but become reality. This book is not all about answers, as after all, in research it is the questions that are the most important, and so there are no detailed answers, and / or massive amounts of supporting evidence, but the questions are certainly worthy investigating, since there will be a need for "different thinking from the one that has gotten us here" to move away from the problems we currently face.

As a writer, one can only hope that "the reader" enjoys reading his creation and benefits from the exercise of doing so. If after all this you are still keen for more, then feel free to "hunt" me down on this planet and follow me through some of the social media, or drop me a line with some of your comments, concerns, etc.. It is after all, always great to hear from your readers.

Acknowledgements

The work presented in this book is based on quite few years of doing research in this area, during which I interacted with many people. Some of this was through one on one meetings, attending seminars and / or group discussions, but there was also the anonymous feedback on publications, and the many sources of information that others published and where I was "the reader". Trying to enumerate all these people would be an endless task, with the risk of someone having "slipped-my-mind", with the necessary but unwanted consequences. So, rather than trying to create and present you with a long list of names, I would like to thank all those that have supported me with their time, discussion, feedback and knowledge through the years in any direct or indirect manner. Now a rule is only a rule when there are exceptions, and the exception that I would like to make is to give some special thanks to someone who has always been willing to comment and discuss my most wildest of ideas, he also ploughed himself through this book when it was not yet publishable/published, and that is: Dr. Greg Cain.

With time being a precious resource, your time-investment as a reader is certainly also much appreciated, and I hope you will find yourself rewarded through reading this book, and watching it stand proudly on your bookshelf.

Considering that professional life is only a certain part of life on this planet, I would also like to take this opportunity to thank all those near and dear to me on a personal level. The support and "believing in ..." is greatly appreciated, especially on those times that I may have been enthusiastic about a rather awkward sounding idea at the most inconvenient of times ... thanks for your patience. As you can see, it is all paying of in the end !

1
Introduction

> *"There is nothing more difficult to take in hand, more perilous to conduct or more uncertain in its success than to take the lead in the introduction of a new order of things."*
>
> – Niccolo Machiavelli

Computing has been around for quite a while, and to many it means the processing of input information towards specific outputs, where the outputs are created through a certain function that fits the user's requirements. The most basic examples of such tasks are: addition, multiplication and so on. However, computing has become much more than that, and to many the underlying methods used are not known, since a computer is simply a "black box", but that does not mean that the expectations for a newly acquired "black box" are not higher than from the previous one. So essentially, functionality needs to expand.

The next chapters will start by providing more details of current computing platforms and how they operate, but the overview will only represent a fraction of what one can possibly call computing. In the broadest sense of the word, there is quite a bit more to its meaning and implementation, and within this context, one should be aware that only a fraction of the possibilities of computing have been explored. As with many things, one tool is good to do a specific task, but often fails to deliver another, and this is no different in the context of computing.

If you are interested in the concepts of what is computing, and what can be computed then [17] is certainly worth your attention. While it debates about which physical system can compute, there are several points made that also lie at the basis of the rest of this book. To set things clearly from the start, the purpose of the book is not to focus on the "inabilities" of current computing, but to clarify its limitations and also the context in which it works properly. Meanwhile, the intention is to indicate how some of the expectations in terms of computing can possibly be met through changes in technology. These new developments do not necessarily have to "take over" from the existing approaches, but it is more likely that multiple systems should co-exist. With "co-exist", I refer to systems that have substantially different strengths, not like the current computing systems described in Chapter 2, which are all based around the same concept.

2
Mainstream Computing

"The mainstream is always under attack."
— Bill Gates

"Once we accept our limits, we go beyond them."
— A. Einstein

Mainstream computing in the context of this book, is considered to be any type of computing that allows processing of data through electronics as currently, at the time of writing, commercially available. In practice, this comprises e.g. general purpose computer, graphic card processors, MPEG en/de-coders, network chips, mobile phone processors, servers, just to name a few. Largely, these all operate in a similar manner and so "how" they achieve "computing" through hardware / software algorithms will be the first topic covered in this chapter, through this discussion the problems faced with these technologies will be highlighted. Further on, the chapter discusses the issue of data-flow within these platforms. The split between "computing" and data-flow, as used within this chapter, has been a common theme in current platforms, and one should consider "computing" to refer to the pure manipulation of data, while "data-flow" refers to the actual "movement" of the data.

2.1 "Computing" in Current Computers

There are several ways one can look at current computing, but the main focus within this book will be the hardware perspective. After all, software is limited by the hardware it runs on, so it is essential to look at the underlying platform to get a better understanding of a system's computational abilities. Various different platforms are available, and so most of the following sub-sections will focus on the main platforms in terms of performing actual computations. However, the very first sub-section will look at the actual production of hardware, and the problems that are currently faced within this context.

2.1.1 VLSI Manufacturing

Ever since the start of VLSI (Very Large Scale Integrated) manufacturing there has been a continuous focus on improving the density of integration, i.e. the number of transistors in a certain area. So far, these improvements have mainly been achieved due to the ability of decreasing the feature sizes of the transistors. By decreasing the feature sizes of the transistor, which practically refers to the width and length of the conductive channel, one can implement more transistors in the same area and consequently produce larger circuits with more transistors. However, the ability to create larger circuits depends on two aspects: firstly, there is the size of the die, which is limited and so determines the number of chips that can be produced per die, and secondly, these dies are never really fault free, and therefore any fault on a die could affect a particular part or even the complete chip. It is intuitively quite easy to understand that if the transistors become smaller the same design requires less space and is therefore less likely to be affected by these faulty areas on the die. On the other hand, if a small fault affects the chip / circuit, then the impact could be larger, since the complete transistor could now be affected and can therefore no longer switch on / off, which stands in contrast to the transistor being only partially affected when its width / length was larger.

Up to now, these shrinking feature sizes have pretty much followed Moore's law, which should probably be called Moore's trend, but in essence says that the number of transistors within the same silicon area will double in about 18-24 months. There would be people arguing that Moore has been the main driver for industry to move forward [3], but eventually it does not depend on who "leads" who, but what the benefits and overall progress are, and especially what the economic benefits are of these developments.

For many years, one of the main advantages of these shrinking feature sizes was the higher performance, since the transistors were smaller and could therefore switch faster. As much as that applied to all VLSI produced devices, this mainly reflected itself in the noticeable improvements of processor speeds in the nineties. It was at that point in time that marketing was "selling" new devices, based on the fact that a higher clock-speed, meant a "better" computer. Rather unfortunately, this image ignored a lot of other factors that contribute to the computer's performance, like the amount of main memory and so on, but that is the topic of a separate discussion. Since early 2000, clock speeds have no longer been changing much, if at all, and so a different route was chosen, namely that of adding more cores, because shrinking feature sizes meant that more transistors could be fitted onto the same die area, and therefore making the chip itself cheaper. Consequently, the marketing strategy changed to making the user believe that more cores meant a "better" computer, once again a rather simplistic view.

As much as shrinking feature sizes led to certain advantages, there are also an equally large associated set of challenges. Firstly, there are the challenges related to the actual manufacturing of devices. In order to manufacture a VLSI chip, one needs to produce several masks which are then used in the lithography process. Shrinking feature sizes has meant that the holes in these masks have reduced in size and while these masks would normally be burned using a light beam, there are limitations to the minimum size that one can create, depending on the wavelength of light. Up to now, methods have been found to overcome these challenges by e.g. using a liquid medium to improve the transfer of light, which overcame problems with the breaking index and so the light can be concentrated more easily to small areas. It is expected that these problems will continue to find solutions, although one can start to wonder what will happen when atomic level sizes will be reached.

Another consequence of the reduced feature sizes is that a transistor and then especially the channel between drain and source, has reduced to such an extend that this distance reaches the size of multiple, all the way down to single atoms. As a result, it has become significantly more challenging to switch the transistor off, since the distance between drain and source is now only a few atoms, and it is obviously not easy for a single atom to either deliver perfect insulation or conductivity. Hence, even when the transistor is in the "off" position, there will be a significant leakage current between drain and source, which leads to an increase in static power consumption. Consequently, several techniques have been introduced to reduce this static power consumption, but most of these techniques introduce extra transistors which are then used to switch off inactive parts of the circuit and only activate them when required. A second consequence of the smaller channel lies in the fact that devices have become more prone to the influences of external radiation. Such radiation can come from a multitude of sources, but in essence any electrons / protons that are flying around can easily affect the behaviour of these small transistors, leading to unexpected behaviour. After all, there are good reasons why mobile phones have, from the start, been banned in certain environments, such as hospitals. Considering that the amount of mobile devices and wireless interfaces are both on the increase, one can easily see that this interference problem is on the rise, and that it would neither be an option to provide the user with a 100+ page manual to explain in which environments and with which proximity the new device can be used in correspondence with existing / other devices, to ensure safe operation.

Considering all these disadvantages, one would wonder if silicon production has a future at all, which is a question with multiple aspects to it. Firstly, one should remember that the majority of VLSI produced chips only has transistors on it. Even the passive elements (Resistor, Capacitor and Inductor) would be produced by exploiting the passive characteristics of active elements. So, if computing and electronics have to remain the way they are, then the new technology would need to provide a replacement for the transistor in this new technology, or a completely new basic element offering similar types of functionalities. Secondly, the production itself could evolve away from CMOS (Complementary Metal-Oxide Semiconductor), and there have certainly been candidates around. However, as much as there may be new technologies around, changes will only take place if the economic model for these new technologies are better than CMOS. After all, the reason why silicon has become as popular as it is, is because it can be produced cheaply. With "cheap" one needs to understand that this refers to the context of "in comparison to" other technologies. One should however be aware that over the more recent years silicon manufacturing has become more expensive, especially when it comes to producing masks, but then again once these are produced, one can make as many devices as required. Nevertheless, the type of devices that are produced in silicon has certainly evolved in comparison to the early days of silicon. While previously, most reasonably sized electronic designs would be produced in silicon, the increased mask costs has meant that, now, only designs with large offset markets are still produced in silicon, and for smaller series, reconfigurable devices and processors have known an increased popularity. Needless to say that this has also had an impact on computing as such, as will be seen later on.

So are there any alternatives to the silicon / semi-conductor industry for manufacturing electronic devices ? Yes, one solution could lie in e.g. using Carbon Nano-tube technology, which has now been researched for a while. However, most alternative technologies look at constructing the same type of basic elements for the underlying / logic layers, namely transistors. The main question then lies in the fact of whether this is the correct approach, considering the above described problems with current electronics. After all, if there are so many problems with the current approach, then will the silicon road continue for much longer or soon appear to be a dead-end street. Maybe it might even be that "electronics" as known now is no longer purely relying on "electrons", but starts to mix with other fields and consequently looks at "chemicals" or other possible "carriers of information". So, as much as there are a variety of different technologies that may or may not be used in the future, there are even more arguments in favour / against the fact that certain technologies may become mainstream or not. However, considering that it is economy and not technology or environment that currently drives businesses, the cheapest option will most likely win the competition. This may change in the future, especially when environmental challenges increase and energy consumption as well as the aspects of a circular economy provide for more stringent requirements, but until then it is still money that does the talking. In any case, for the time being, development still seems to be ensured and solutions are still abundant, so, one can still sleep on both ears, since new gadgets will continue to reach the market.

Coming back to the decreasing features sizes, this has not only been driven by Moore's law, but to a large extend also by digital electronics. Considering that digital electronics makes use of an abstraction between the underlying transistors and the actual logic levels used within the circuitry, there are a larger number of transistors required to provide for a quite simple function (AND, OR, etc.) from a "computational" perspective. Consequently, the ability of adding more transistors onto a single chip has provided massive benefits in terms of producing more complicated digital electronic circuits. On the other hand, analogue electronics exploit the full characteristics of every single transistor and in this context it is better if the transistor's characteristics are less influenced by other factors, which is the case if the transistors become smaller, and so analogue designers generally prefer larger feature sizes. However, with increasing miniaturisation and the demand to decrease energy consumption, the boundaries between analogue and digital are decreasing, and techniques from either are used in "the other field" to overcome some of the problems faced, which is also where possibly the long term future for electronics lies. So, the separation of analogue and digital is likely to disappear to make place for mixed designs.

While shrinking feature sizes have challenged transistor manufacturing, they have also allowed for other / new types of basic elements / components to become available. In the early 70's, Leon Chua [2] argued that there was a missing passive element. While, most people are familiar with the following passive elements: resistors, capacitors and inductors, who describe various relationships between the variables: current, voltage, flux and charge, as shown in Figure 2.1, one can notice there is a missing relationship within this puzzle, namely the relationship between flux and charge. L. Chua proposed that this missing element would be a memristor and have a resistive value with the ability to retain it, i.e. memory functionality. The element remained hypothetical, until HP Labs identified that one of the items they had produced on silicon at nano-scale, was actually a memristor [8]. The

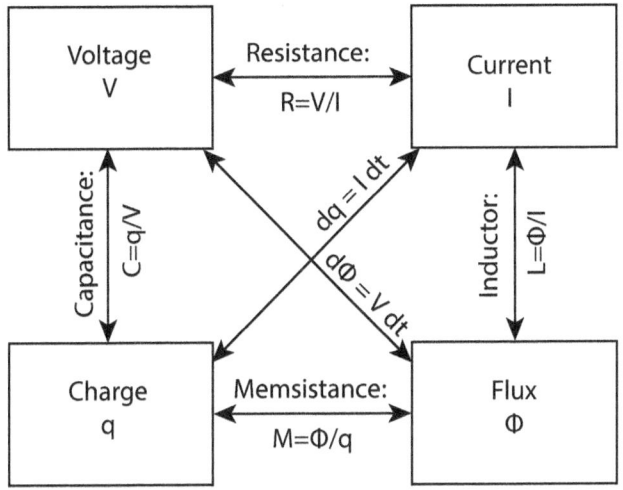

Figure 2.1: Overview of Passive Elements

interesting aspect to this discovery lies in the fact that this element only behaves like a memristor when produced at nano-scale. The fact that one was able to produce this passive element could potentially mean a massive shift for electronic development. Considering that it is a passive element, means that the power consumption will be limited, but it being a device with memory effect, also means that it will retain its state when there is no power, which could e.g. help for instant-power-on applications, and these are only a few of the potential benefits.

2.1.2 Application Specific Integrated Circuits

Application Specific Integrated Circuits (ASICs) are currently mainly used for designs that either have a large offset market and / or a very specialised application domain. The latter could for example be: aircrafts or military, where money is generally less of a constraint, and reliability and continuous operation are key requirements. The more common area of ASIC development would however focus on chips with specific functions and large offset markets, like network communication (wired or wireless), MPEG 2/4 decoding for DVD players and so on. Any required "computing" within these devices is obviously hard-wired into the silicon. The algorithm that ensures the appropriate processing is likely to have been translated from a mathematical and / or software description to be implemented onto actual hardware, which could entail simplifications, such as using fixed point instead of floating point and so on.

Most of the translation from algorithm description to actual hardware description is done using hardware description languages, and requires a large amount of human interaction, especially when optimal performance is crucial. While synthesis tools are good at translating very generic code, the more the designer helps, the better the outcome will be on actual hardware [14]. One can then still perform further optimisations if required. For example, during the design of general purpose processors, transistors are all sized individually, to achieve optimal performance, and while this is a very labour intensive process, it is the only way to achieve such high performances. The general trend is however to move towards higher level description languages, where possible. Synthesis tools are then used to generate the actual hardware, and while the level from which these synthesis tools start, improves to more abstract languages, the disadvantage is that the outcome is generally less optimal when compared with a design that would have been created using a lower level language. This also means that these higher level description tools are generally more useful for quick "reality" checks rather than actual high specification designs. One of the main reasons for this lies in the fact that it is particularly difficult to automatically optimise data movement.

With time-to-market ever decreasing, the designing of ASIC devices has also moved more towards the use of custom blocks. One of the most popular examples in this context is certainly the ARM processor and its use in e.g. mobile phone chips. While this helps with reducing the design time, it does not reduce the costs. Consequently, if performance, among other can be compromised then it may be better to use reconfigurable or general purpose processors, which provide hardware or software flexibility respectively.

2.1.3 Reconfigurable Hardware

Somewhere in the middle of the full spectrum going from Full Custom / ASIC all the way to processors lies reconfigurable hardware, who is gaining terrain from both sides. There are a variety of reason for that, not the least because they offer a flexibility similar to that of a processor with the performance closer to that of full custom designs. The perfect wedding, one could assume, although what may seem to be perfect at first sight is not that perfect. After all, the down-sides to reconfigurable platforms are their power requirements as well as the significant performance problems which, for the majority, are due to their routing infrastructure. As much as these problems are being worked upon and have improved to some extend, since they are related to the underlying architecture, they are unlikely to completely vanish, unless the flexibility of reconfigurable devices is being traded in.

While reconfigurable devices have become the preferred platform for quite a few applications, they have not quite managed to meet the initial expectations of forming replacements for general purpose processors. One of the main challenges towards reaching this goal lies in the long lack of on-the-fly reconfigurability. Most devices now offer a solution to this through partial, on-the-fly reconfiguration. Hence, this has certainly helped to make these devices more flexible, and allows e.g. smaller devices to be purchased for certain applications where partial reconfiguration is possible within the context of the application, because not all tasks need to run simultaneously. However, there are still certain steps to be taken, before the technology will be able to be used in areas where processors are now more beneficial.

The increasing costs to manufacture full custom devices, have certainly helped the move towards reconfigurable hardware, since they also fill a gap from an economic perspective. After all, they can be produced in large quantities and are therefore reasonably cheap devices. Considering that the design methodology to be used for creating the actual designs is in no way different from that for full custom designs, designers do not have to be re-trained either. However, with the increased use and sizes of these reconfigurable devices, the need to improve performance has become more prominent and has actually been met through the addition of flexible custom-function blocks. These custom blocks have then mainly been designed to deal with some of the application requirements for which these devices are quite often used, e.g. the multiply-add requirements for FIR (Finite Impulse Response) filters in DSP (Digital Signal Processing) applications. For all other operations, flexible logic, in the form of Look-Up-Tables (LUTs) is used. This approach has the disadvantage of lower performance and higher power consumption in comparison to full custom implementations, because there is a need for extra switches and storage logic, which require power and area. So, reconfigurable devices have to get their performance benefit over processors through the exploitation of parallelism. However, not all algorithms can exploit large levels of parallelism, and it may require human intervention to ensure the optimal utilisation of the available resources.

While most of the above applies to so-called fine grain (i.e. LUT based) reconfigurable architectures, there have also been developments towards more coarse grained architectures. The latter would then have a basic cell with the ability to e.g. perform 4-bit processing, with a small instruction set. While coarse grained architectures can bring performance benefits, they would certainly also have their limited set of applications to which they are beneficial, due to their "lower" flexibility.

Designs made for reconfigurable platforms are generally described in a higher level language and then mapped onto the actual hardware, similar as to full custom designs. The main difference however lies in the fact that in comparison to full custom design the tool does not have a "blank sheet" as target, but a specific underlying architecture. While a substantial amount of work has been done in this area, as with any translations, there are a set of limitations and quite often better results are obtained from one language / description than from another. Any designer with experience is aware that these synthesis tools have allowed for shorter development times, but not necessarily result in a design with optimal performance. Achieving optimal performance can still be reached most easily by taking the underlying platform into account when doing the design work, which in essence eases the task for the synthesis tool itself [14].

2.1.4 General Purpose Processors

Processors have been around for a long time, resulting in a large set of processors and suitable computer architectures having been proposed and used in a research and / or commercial setting, each providing their respective benefits. Consequently, a large variety of different types of processors are available, with many differences between them. The largest differences lie in the instruction set and, as a result of this, also the application domain, for example: general purpose processors, graphical processors, embedded processors and so on. As much as there may be differences in the instructions that particular processors support, there are also differences in how they deal with data. This would reflect itself in being able to deal with one or multiple data items during the same instruction cycle, where the latter would then generally be referred to as vector processors.

While general purpose processors can be used for a large set of tasks, being able to deal with more tasks, also means they are less focused on one specific task, and so less efficient for very specific tasks.

Ideally, the instruction set is as large as possible, but there is a trade-off between the number of instructions, which directly affects the size of the processor on silicon, and the efficiency of operation. Additionally, extensive instruction sets also complicate the design of compilers, which take a programme designed in a higher level language and translate it to the language that the processor will understand, namely its instruction set. Any tests on compilers and processors show that if the compiler can make full use of the available instructions, then the performance will be more optimal [10], but the compiler's task is far from easy, since quite often there is a large set of potential solutions and so the true challenge lies in determining the optimal solution in the shortest possible amount of time. As you may start to appreciate, computing in its current form is getting more and more complicated, not to say that there is a lot of inefficiencies in some of the processes, especially when it comes to algorithms themselves. Algorithms are used to solve a variety of different problems, e.g. object recognition for autonomous driving. Even for such an application, there would be several different algorithms available. Each of which would be slightly different in exploiting the benefits of the platform on which they run, and that to achieve the goal of object recognition. Usually, this also means there is a trade-off between execution time and accuracy for these different algorithms. New algorithms, or improvements to existing ones, tend to shift this balance more one way or the other, but rarely improve both factors. However, a lot of these algorithms take a long, and indirect way to their destination, since they perform a lot of calculations on a typical number-crunching processor to lead to one single result that is then used for further processing. The important question is whether this is the shortest and most efficient path to the destination, or whether shorter paths are possible, and if so how can these be exploited.

Coming back to the previously mentioned improvements to processors, namely the increased clock speed, followed by the expansion on the number of processor cores. Increasing the clock speed improves the processing outcome because tasks are executed much faster one after another, which is particularly beneficial if the tasks are purely sequential. The addition of extra cores is however only useful if tasks can be executed in parallel. However, sometimes less is more, and that is especially the case when there is no parallelism available. After all, if tasks are purely sequential then one would need multi-tasking and / or multiple software programmes to "entertain" each one of these multiple processor cores. However, if only one task is running, then it is not particularly easy to split that task to run on multiple cores. Such splitting could have been done when the actual code is written, which is often the better approach, since compilers and Operating Systems are not always the best at splitting tasks. If you consider that most software is written for sequential execution, then you may intuitively understand that trying to find tasks that can be executed, in parallel, may not be that easy. Secondly, one should also take into account that while the human brain is massively parallel, we are better at designing code for sequential execution than parallel execution. When many tasks are running in parallel, like in actual hardware, then the brain tends to struggle with building up a picture of what happens simultaneously. So, in essence, while multi-core machines have been well marketed, they have not really brought us the benefits they could bring us. So considering that the scaling of processors has looked at more instructions, faster clock speeds and that we are not really benefiting that much from more cores, the question then becomes: What should be the next move to improve computing ? Would it lie in going back to basics and consider efficiency and scalability from the start ?

2.1.5 Heterogeneous Computing

While the previous sections have provided details of various different hardware platforms, there is a trend to start combining these into one larger platform, generally referred to as a heterogeneous computer. The true benefits here are that one can combine any set of technologies as required by the target application, however with such a vast set of possible combinations the biggest question lies in identifying the right balance / combination. So, while this approach allows the design to be optimised for a variety of specific applications, it is quite likely that such an architecture will only truly "fit" one application well and so a large set of solutions would need to be created. Not to mention that this would diversify the solution field and while currently more general solutions can be used, the diversification would only complicate the design and use, which is probably also one of the reasons why this has not really taken off. However, this again does not solve the problem as to how one can design better processing platforms, and then there is still more to computing than only the processing of the data itself.

2.2 Data-Flow within Existing Platforms

While a lot of research and development has focused on processing, currently, of the time that a designer spends on developing a product, the majority will focus on finding an answer to the question: "How do I get the data to the right place at the right point in time ?". The best solution to this answer depends on a variety of parameters of the targeted platform, such as: memory access speed, type of access performed (purely random or with a predefined pattern), available amount and types of memory, as well as the available communication interfaces.

In current designs, communication, memory and processing are all treated as separate parts. Now, in essence, communication and memory are not really that different, because communication means that the data is moving with a certain speed, whereas memory means the data is steady for a particular amount of time. So, in essence, the difference between communication and memory lies in the relative speed of the data, which would be zero for memory and a number relating to the communication speed of the particular medium, for communication itself. So, within the context of this book, communication and memory will be treated as one and referred to as "data-flow", where the speed of the "flow" refers to that of relative data movement.

The largest problem with data-flow as it stands currently, lies in the fact that it has not developed at an equal rate in comparison to processing logic. This is what generically is referred to as the Von Neumann bottleneck, whereas Von Neumann actually did say that processing logic and memory should be developed at an equal pace, because an imbalance between the two would lead to problems. Obviously, one could then wonder, if both would have been developed at the same pace, would the current problems then still exist, or are we just looking for someone / something to blame ? So, in essence if we manage to bring data-flow development back in line with what we achieved for processing, through the development of more and faster memories, then would there still be a problem ? Personally, I think yes, since those are not the only problems with our computing model. So, the problems are bigger.

To highlight the significance of the Von Neumann bottleneck in current architectures, an analogy will be introduced. Assume that you have a team of people that need to do a particular task (processor), and this requires a certain physical element (data) to be delivered from stock / store (memory). So, for example you want a team of people to paint eggs for Easter, and you will tell them which patterns and colours to paint on the eggs, then if any person wants to start painting, according to our current computing model, this person would first need to ask for an egg and then also ask for details about the required colours and pattern he / she would need to paint. This would then be supplied to him / her from stock. A second person would need to do the same and so on. All would be fine if they put in their requests one after another, but imagine that they all need to paint as many eggs as possible (optimal performance), then they would all want to start together and would consequently all put in their request at the same point in time. At that point in time, the person who deals with the requests at the stock side will struggle. The best that he / she can do is to deal with the request one after the other. So assume that this is what happens and so the first processor can start working. One can additionally assume that all processors are doing the same task, and are equally qualified, so would take the same amount of time to finish their task. If they need to provide the finished egg back to stock when they completed their task, then they will need to do so before they can request a new egg and painting instructions. Meanwhile, the person in stock would have been twiddling his / her thumbs while the people / processors were working. Now, if this person would know what is coming, he / she could prepare the next delivery for each of the processors. This would obviously require the person to know which pattern of requests there will be. As it has been shown over and over, these patterns do exist, since there is spatial and / or temporal locality in data being requested from memory for most computing applications. This is exploited to a certain degree, in that when data is collected from a slower memory then the "surrounding data" would also be collected and brought into the faster memory, and so we can build a memory hierarchy with smaller and faster stocks at the top and larger, slower stocks at the bottom. In terms of our analogy, this would mean that multiple eggs and patterns are moved to a faster stock / store. This however does not eliminate the principle that everything needs to be requested by "the processor", which is one of the problems with the model itself. As much as this cannot be easily overcome in all environments, if the eggs and patterns would be supplied to the people in time, according to the JIT (Just-In-Time) principle then that would be the most optimal situation. To achieve this, one needs to know the access pattern on one hand, and on the other hand the store needs to have sufficient "intelligence" to deal with finding the items, since items are generally stored in locations with a numeric address.

So, while most of the current computing models are based on a data-request principle, other approaches have been considered, such as stream based processing. This has the advantage that data is provided continuously at a certain rate and passes through the processing logic. As a principle this can work nicely for certain applications, such as video processing. Although that requires the data to be processed in a certain order, as soon as multiple data items are required, or different data orders are needed, then this model is less effective or no longer usable.

In the same line, one could argue, that rather than focusing on the processing side, one should focus on data-flow, which is what happened in more memory centric architectures, but since they did not overcome the problem, they did not develop further. One such an example is the PIM (Processing In Memory) devices [4, 5], which bring processing and memory much closer together, with the purpose of reducing the bottleneck. This approach brings along a number of design problems, such as: what would be the optimal ratio of memory versus processing logic; how complex should the processor design be in relation to the available memory for that processor, and obviously the answer to these questions are: it depends. Indeed it depends on the type of application you are dealing with, and so quite again it is impossible to create a single design that works for all applications. So, are there any solutions ? Yes, but before those are being looked at, it is essential to get a better understanding of the technologies that ensure data-flow and their implementation within the current architectures, since communication will be key.

2.2.1 Hardware Implementation

Considering that communication and memory are generally developed separately, this section will treat them separately to represent the current situation. In terms of communication technologies, the two main types are either bus or network based communication and then one could either have permanent connections or package based communication between source and destination. The permanent communication could then still make use of parallel or serial protocols. As with all technologies, there are respective advantages and disadvantages to each one of these technologies and there are certain cycles of popularity for each one of them. After all, serial (e.g.: serial port) was replaced by parallel (e.g.: parallel port and PCI) to again be replaced by serial (e.g.: PCI Express), and then in some cases again combined with some of the principles of parallel (e.g.: PCI-Express x16).

In terms of memory, the challenge has always been to find a fast and cheap memory that preferably retains its state when power is taken away (i.e. non volatile). Meanwhile, a memory hierarchy, with expensive, fast, but limited capacity memory is found closest to the processing logic, while large, slow and inexpensive memory is found at the bottom of the hierarchy. In between, there could be any number of stages. The challenge with this model obviously reflects itself in moving the data in between the different levels of the hierarchy to optimise access performance. This can easily be achieved when the access pattern is known, but is quite a different challenge if it is not.

The number of applications with a known access pattern are however limited, so with random access being more common, the fast provision of data to the processing logic remains a true challenge.

When looking at the actual hardware implementation of communication, then it is no surprise that certain systems suffer due to the limited number of communication channels. Ideally, the communication bandwidth would be infinite, or at least from the perspective of the application context for which it is used. However, infinity is not essential to make a system work, but considering the data request model in current computers, there is certainly a requirement for more data bandwidth to deal with the two way communication requirement. As you will have understood by now, when designing a system, there are a large number of factors to be taken into account, namely: processing, communication and memory, but also the hardware and software sides are to be considered to ensure that the platform on which the algorithm runs is as efficient as possible. From this perspective, and based on the similarity between communication and memory, one could wonder if there is a need to separate: processing, communication and memory. Taking that the latter two form data-flow, why would it not be possible to also "process" this data, while it is being moved. After all, if one looks at e.g. the human brain, which could be considered as a data-flow centric architecture, then that seems to be exactly what is happening. However, there are still a number of challenges to be overcome before a human brain can be implemented in VLSI.

2.2.2 Application Specific Integrated Circuit

In ASIC architectures, the data-flow part would be custom designed for the application in question. This allows for the processing logic, memory (temporary storage) and communication to be well balanced. However, any balance also has its trade-off and these then relate to parameters such as area and performance. One important thing to keep in mind is that most design techniques used in this context are generally borrowed from the ones used in computer architectures, especially since a lot of the standard cells used during designs are based on these principles. So, while data-flow in ASIC devices is designed for optimal performance, it still tends to come secondary to the processing design.

In terms of silicon implementation, the transistors are created on a 2-D wafer of silicon, and then the layers above these transistors provide for the interconnections through a variety of metal layers. With process improvements, the number of metal layers has increased, since there is obviously a relationship between the amount of transistors that are implemented on silicon and the required number of metal layers for interconnection. When one then takes into account that on-chip connections are significantly faster than connections over a PCB, then it is fairly straight-forward to understand that one prefers to put as much as possible on a single chip. However, long metal on-chip connections, suffer from capacitive and resistive effects due to them being small and thin metal conductors. These effects would obviously influence fast communication requirements between e.g. memory and processing resources. Therefore, one of the next developments that may be commercialised is the use of a 3-D chip construction. As much as this is referred to as a 3-D structure, it is in essence a stacking of two 2-D structures, and therefore more like 2.5-D. The difference with true 3-D is significant, because for true 3-D effectively any point can be connected to any point using a direct path. When stacking 2-D structures, the number of interconnections are limited by the stacking used and therefore suffer from certain limitations. After all, communication requires communication paths, in electronics, wires, and the more wires one can have, the more directions one can communicate with between different sources / destinations, but obviously wires cannot collide or they would be "sharing" information. So, while these new 3-D structures will provide an improvement to the communication possibilities, the question is whether there will not be another boundary around the corner, and so whether communication and other problems should not be dealt with at a more fundamental level.

2.2.3 Reconfigurable Architectures

Considering the need for flexibility in reconfigurable platforms, this does not only require flexibility in the processing itself, but even more important is the required flexibility of the on-chip connections. This flexibility allows for connecting certain lines to the inputs / outputs of internal blocks to provide the required functionality. To achieve this flexibility switches / transistors are used, which introduce a delay along the path. Additionally, unless the software manages to keep all processing and communication local, wires with less switches along them are being used, to limit the time delay introduced by these switches.

When memory is added into the picture, then it becomes even

more challenging, since the amount of on-chip memory is generally limited. The challenge therefore lies in determining which data should be brought on chip and where it should be stored, to ensure it is as close as possible to the processing logic. Again, the more optimal results are achieved through human interaction / development, since these tasks are particularly hard for automation tools. Designing could be made easier through the provision of high level abstractions integrated in the design language. However, also here it quite often turns out that the use of a hierarchical memory design is most beneficial.

The underlying architecture of reconfigurable devices, of the type most regularly used in industry, are actually build upon a basic cell that is no different from a tiny memory, generally referred to as LUT (Look-Up Table). Logic processing is performed by using this LUT as a truth table otherwise well-known in digital design. Obviously, one can also use the block as a tiny memory although that is rarely very effective due to the small storage capacity. However, coming back to the differences between processor, memory and communication, this principle clearly indicates that processing can be done through memory. One could and should wonder whether that is the most optimal way of processing, since that is therefore not a guarantee, but at least it shows that one does not really need many different types of logic to offer the different "functionalities" as currently known and offered in computing. The question then is whether this will allow for new developments, and then one has to take into account that one of the challenges with doing such developments is that people's vision is generally restricted to what they have learned / seen before. After all, it is said that when America was discovered the local population had not seen the boats coming because they had never seen a boat, and this is no different.

While FPGA platforms have a reasonable amount of memory available close to the actual processing logic, most applications mapped to an FPGA also show regular access patterns. Hence, accessing the memory could be automated, by expanding the memory with logic that autonomously provides data to the processing logic [11, 12]. The only thing the processor has to do then, is indicate that it wants to receive new / more data or is too busy. Additionally, if the computation time for a particular item is known, then one can even provide the next data in time to the processing logic, which in concept comes very close to stream processing, where a data stream is passed through some computational logic, but in this case the stream would be customised. Adding autonomous memory blocks to reconfigurable devices brings substantial benefits to the FPGA architecture [11]. It does not only make more efficient use of the silicon in terms of area, but it also allows for designs to be run at much higher frequencies. Such autonomous memory blocks can also be combined [12], which allows for storage of any size of data while still providing the same autonomous functionality, and retaining the benefits obtained from a single Autonomous Memory Block (AMB).

In a next step, the principle of autonomy will be taken one step further. So, how about if there would be small, completely autonomous cells [14], where autonomy is inherent in the processing as well as the memory storage. Essentially, data is automatically provided to the chip, but also automatically provided to the processing elements within the chip. This architecture is then compared with an architecture where everything evolves around a processor, leading to the results shown in Table 2.1. Each time that a task that leads to the processor being the bottleneck, this task is taken out of the processor and implemented in actual hardware. Over the course of doing this, one can see the area increasing and the performance improve slightly, but the results are still no where near the pure cellular approach. While, the cellular approach has the additional benefit of being perfectly modular and scalable. This proofs that using a different design concept can significantly change the features of the design, even if the same underlying architecture is used. So, there is a huge responsibility given to the designer in terms of achieving optimal performance, and wrong decisions in this context can carry severe consequences. After all, the translations tools, in the form of compilers and or synthesis tools, are only able to optimise within certain boundaries, and obviously the better the input design is, the better the output design will be. Ideally, one would however prefer to remove this dependence on human intelligence and manage to create fast applications independent of how they were designed. Certainly an interesting challenge, especially considering there is a variety of levels that can be used for the design phase, and an interesting set of challenges to be overcome in going from problem description over algorithm development to actual hardware design.

2.2.4 General Purpose Computer

In most cases, the processing model in a general purpose computer is based around the Von Neumann architecture, which has a central processing unit, with two separate units, namely one for control and one for arithmetic / logic functionality. Furthermore, there is a mem-

Table 2.1: Performance and Area for Heterogeneous Implementation (Search-Size ± 4)

Implemented architecture	Max Clock Rate (MHz)	Area (# LE)	Memory (Bits)	Processing time per macroblock (msec)	Throughput per LE
NIOS II only	50	2411	46592	50	0.21
NIOS II + AD unit	50	2547	46592	39	0.25
NIOS II + AD + AAC	50	3605	101376	22	0.31
NIOS II + AD + AAC + External Memory Controller	50	5099	101376	23	0.21

ory unit which stores instructions and data, and a number of input / output devices that can be linked up to this general architecture. This model is best known for the Von Neumann bottleneck, which in essence refers to the limited bandwidth between processor and memory, resulting in the processor regularly awaiting data from memory. A slightly different model is the Harvard model, which separates instructions from data, but in essence is also a request based model where the processor is fully in charge and requests instructions and data from the memory as and when required. So, while in each case the processor needs to put out requests, if the memory who needs to deliver the data is much slower, then the processor is twiddling its thumbs awaiting to receive this information, and therefore overall efficiency suffers.

So, one could argue that the processor needs to slow down, but then the only way to get more done is by using parallelism, which then again multiplies the number of simultaneous requests in line with the amount of parallelism introduced and therefore not really eases the memory's function. Secondly, one could look at improving memory, making it faster and / or more intelligent / independent. In terms of memory, a large range of developments have already taken place, from memory for short and long term storage over to special types like: CAM (Content Addressable Memory), Associative Memory and so on. There have also been a few designs that introduce intelligence within the memory, since generally the memory only performs a storage function combined with the ability to retrieve the stored data. Since memories temporarily store data / information they also need to retrieve what was previously stored. This retrieving is performed through using the address which then returns the data or, in case of CAM or associative memory, by providing data and then returning the address location at which the data is stored or the closest matching data respectively. However, neither of these designs improve the response speed of the memory, it is only through the development of better / faster memory that one can retrieve multiple items per clock cycle, like e.g. DDR2, DDR3, where the number of responses per clock cycle has been increased.

The Von Neumann bottleneck is however not only applicable to general purpose processors. Different types of processors, such as: DSP (Digital Signal Processing Unit), GPU (Graphical Processing Unit), etc. each target a specific function, but in essence work in a similar way as the general purpose computer, and are all based on the data / instruction-request model. The larger the amount of parallelism used, the larger the access bandwidth requirement to memory. Systems with more parallelism are nothing new, since they have been around in e.g. large data centres, but now also found their way to desktops and mobile computing. So, the bandwidth problem has been around for a while, and there have not really been any working solutions so far, which makes one wonder whether the current computer architecture, is not secretly bumping into a wall that will not budge, meaning that the only solution will be true a step change.

3

Intelligent Hardware

"Failure is simply the opportunity to begin again, this time more intelligently."

– Henry Ford

So, having looked at the hardware problems related to current architectures, the main question is whether there is a platform that can overcome these problems and if so, how can it do that while progressing computing towards a brighter and more capable future. More capable would then most likely refer to reaching the "abilities" of the human brain, more generally referred to as intelligence.

With "intelligence" being a rather generic word, it is important to first clarify what it means within the computing context and also why there is a need for it. The chapter will then continue to explain some of the requirements and how these could be reached, or how such a platform can be designed.

3.1 What is "Intelligent" Computing

First things first, using the word "intelligence" can spark a long discussion as to what it actually means, especially in a computing context. Most people would probably even argue that current computing platforms are already intelligent in one way or another, and to some extend they are, but not to the true, full meaning of intelligence. As much as there has been a lot of design effort going into improving current architectures / platforms, and they are becoming more and more "intelligent" in the sense of e.g. energy saving, autonomous behaviour and so on, this is not the true and full meaning of intelligence. One could probably classify the existing levels of intelligence as "partial intelligence", because of the fact that the intelligence is limited to certain domains of operation, but rather than spending much time on trying to define the intelligence of current computers, let's look at what "truly intelligent computing" would really mean.

To do this, it may be best to start looking at the meaning of the word intelligence itself. Intelligence comes from the late Middle English, via Old French from Latin *intelligentia* from *intelligere* 'understand'. So, while originally it seems to link with the more passive ability to 'understand' most would probably be more familiar with its definition as: "The ability to acquire and apply knowledge and skills", which is no longer passive, but quite active, as it allows one to apply. Now, applying this in a computing context, one would obviously end up in the field of: Artificial Intelligence (AI), which in essence refers to machines and / or software that exhibit "intelligence". The latter use of intelligence again tends to go back to the one of our previous paragraph, but essentially AI seems to refer to the fact that machines have the ability to show / exhibit such intelligence. However, this book will still take the definition of intelligence a bit further.

After all, in relation to AI, it is fairly easy to say that a machine shows "intelligent behaviour", without really looking at how that machines achieves it. Considering that AI has been around as an active research area for quite some time, and while some "intelligence" has been coming to our devices, these AI machines are still far from achieving intelligence as known in the context of human intelligence. So, should "intelligence" in AI have a different meaning from its standard meaning. Obviously, that was never the purpose, since then the only difference would lie in it being Artificial. The reason for the current limitations within Artificial intelligence will be discussed in more detail later on, but in the context of this book, the aim is to mimic human intelligence, as provided by the human brain, and so while in theory this falls under AI, for the time being, this will be referred to as intelligence.

There are surely many other definitions of intelligence, and one can even raise the question as to how one should measure intelligence, but while on one hand this is a very abstract topic, on the other it would probably bring us to the essential creation of some type of scale. While one could argue to something similar to IQ, but then for computers, at this stage, I would consider this not to be the most prominent question to be answered. A bit of vague abstract thinking does, after all, not hurt. The challenge of defining intelligence in a computing context is however not new and lies at the source of the so-called Turing test. The test consists of having person A interact with a device / system

/ person which is on side B. If person A cannot make up whether the "item" on side B is a computer or a person, then one can conclude that the computer shows similar intelligence to that of a human. As you may appreciate, this is a particularly challenging task, and other from having developed computers that can beat a human at chess, there have been no applications so far where the computer could pass this test. Additionally, this also shows that Turing was already thinking about machines of the future.

3.2 The Need for Intelligent Computing

As from the very start, computers have been used to make certain tasks easier. These tasks originally focussed mainly on the performing of calculations and / or the quick retrieval of vast amounts of data that needed to be retained and accessed easily and quickly. The latter obviously relates to large databases that used to be paper based and have been replaced by their electronic counterparts. One of the best known examples of an electronic database is probably the internet, which is in essence a database of databases. So, up to now, the uptake of computers is largely due to the fact that they make things easier, because they make information more easily accessible, but they also help us humans by elevating us from certain tasks. In a similar trend, that is also one of the main reasons why a growing number of appliances start to be driven / controlled by electronics, since this way these devices can deal with "more complicated" tasks, such as e.g. making coffee by the time you wake up, due to a clock being built into your coffee machine. As a next step one would expect to be able to programme the machine to prepare a specific type of coffee at a specific point of time of the day and / or day of the week. In any of these cases, these tasks would still have been pre-programmed by the user, since our coffee machines are still not able to know what the user wants unless he / she has programmed the machine with this knowledge. However, this approach is still far from true intelligence, since all information has been pre-programmed.

In the light of continuously improving technology and allowing it to help humans in their daily tasks, the true question is what would be the expectations of the "next generation" of coffee machines. These new items will then also ensure an economical drive to sell because people want to buy the latest and greatest. However, new products also need to offer more, or one cannot quite talk about progress as such. In the context of the coffee maker, this could mean that the coffee machine can produce coffee by e.g. receiving a message from another networked device and consequently produce the required type of coffee by the time you arrive at the machine. This could e.g. be useful in large coffee-shops, where one currently goes to the till to order and then collect his / her item, since this would allow for the till and the coffee machine to be linked with one another and automate some of the steps now performed by humans. One could envision this to be of similar usefulness in office and home environments. Nevertheless, this still classifies more as automation than intelligence. Taking it one step further, would then be a machine that learns from user habits. E.g. there may be cases where if you have a meeting with a particular person you would need a strong coffee either before or after the event to either wake up and / or get yourself back together. This is generally where fellow humans, e.g. secretaries, have proven to be worth their weight in gold, because they know the habits of the people they work for / with and can offer him / her this kind of a service, generally based on previously acquired / learned information. However, with salaries increasing and an ageing population, there is very likely to be a market for devices who can deliver a more customised solution to the user.

By now, you probably wonder, but that is what AI and NN (Neural Networks) are doing, so the solutions are already available. Well, not quite, since these technologies are not fully ready for commercialisation. Without going into the details of what market "ready" means and how it can be achieved, it is probably best to look at where these technologies have gotten too in the time that they have been around, and those achievements are rather disappointing. So, the question then is why did they not "break through" ? For which there is probably a multitude of reasons, but let's focus on what are probably some of the main ones. Neural networks try to artificially mimic the neuron structure of the human brain, but, up to now, the functioning of the brain is still far from fully understood, and it is probably not difficult to appreciate that it can be challenging to build a system that works identical to one that is not fully understood. Secondly, there is not yet something like a truly artificial neuron, although this is being worked upon. However, this also means that all neural networks are simulated using existing hardware, which is certainly not ideal for the job. While these existing platforms may be able to model the electronic part of an actual neuron, real neurons do not only "communicate" with electrical signals, but they for example also use a set of chemicals to exchange a multitude of information. Therefore, the implementation on a (digital) electronic platform immediately limits the set of values and representations in correspondence with reality. Therefore, one of the concluding questions could be whether it makes sense to build up the behaviour from this low level, in comparison to looking at the functional behaviour of the brain (as described in [6, 21]) and trying to provide this functionality using available technology. After all, for many decades algorithm development has been doing exactly this, so why would one not be able to recreate the functionality of the human brain.

One of these approaches looking at providing human brain functionality, is called: Hierarchical Temporal Memory (HTM) [7]. As originally designed by Numenta the HTM algorithm had an underlying model that was originally, largely built around statistics. To the user, the design was pretty much a black box, but in order to use it, a number of parameters needed to be decided upon, e.g. the number of layers within the model. Each layer consists of cells which have two parts, where the overall structure has sensory input on one side and a type of category output on the other as shown in Figure 3.1 a). The two parts of every cell directly relate to the two main functional concepts of the human brain, namely: spatial and temporal locality (See Figure 3.1 b)). The easiest way to explain these is through means of an example. In case of spatial locality, one needs to imagine a graphical example in which an image is sub-divided into pixels. There will be certain lines, corners, etc. that allow us humans to identify objects within such an image, and hence the HTM network stores information about the location of these lines, corners and so on as shown in Figure 3.2. Temporal locality is best understood as the sequence in

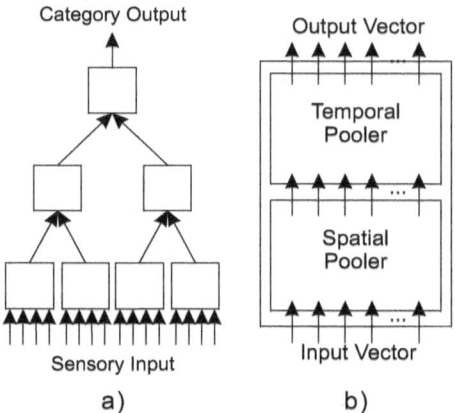

Figure 3.1: Overview of General Hierarchical Temporal Memory Network (a) and each of the Specific Nodes (b).

which information is presented. In terms of visual information, that would relate to e.g. a video sequence, which is no different from a sequence of pictures and then one could ask the user what the next picture could be, or what is being recognised as happening within the sequence. Another example of temporal locality can be found in audio where playing the first few notes of a song does not allow one to recognise the actual song, but a short sequence of notes allows one to recognise the song and even start to sing along. Needless to say that the main power of HTM lies in the data provided during the learning, which is also where a lot of work would still need to be performed. A proper learning method would enable the system to have the correct knowledge to also deal with prediction, based on having been learned a more complete sequence.

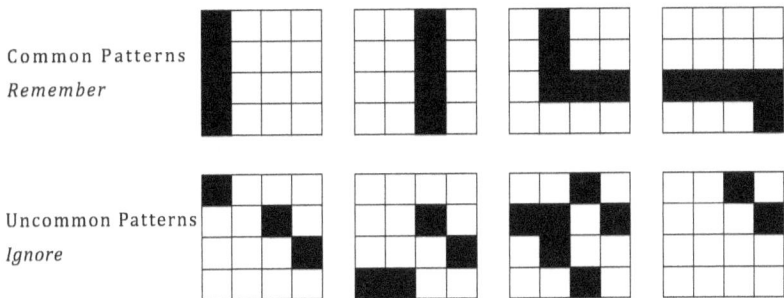

Figure 3.2: Patterns that would be Stored / Ignored for Storage in the Spatial Pooler

HTM has, similar to other AI algorithms, proven to provide a certain intelligence. For example: the HTM algorithm has been used to perform predictions in relation to large sets of data, in which previously no specific patterns could be identified. The reason for HTM finding these patterns is because most previous approaches only considered patterns in the spatial domain, while the temporal domain was not considered at all. Since the original algorithm was presented by Numenta, it has been redeveloped with the focus of using it for large data sets, which has enabled HTM to predict e.g. market trends, in addition to its ability of determining the difference between e.g. cat and duck images.

Gathering from this experience, there is certainly a market for more intelligent computing devices, which can be used in a variety of situations, including e.g. robotics, autonomous vehicles and the like, just to name a few. This brings us close to the question / fear, shared by many, on whether computers will become so intelligent that one day they will take over the world. While this could be considered to be more of an ethical question, the true nature of the question lies in the technical abilities that one can achieve and their threat to humanity. It is obviously a thin line between what people want to achieve in terms of intelligence and what would be considered acceptable without becoming a threat. On top of that, one can never fully prevent the use of technology for malicious purposes, even if that was never the intention. For example when 3D printers were constructed they were never intended to be used to print weapons. So, how "close" could technology bring us to having machines rule the world ? Well, first of all it would require us humans to fully clone the human brain with e.g. electronics, and as mentioned earlier, electronics on their own are unlikely to achieve this goal. Secondly, technologies like HTM only provide certain functionalities of the human brain, namely the prediction / similarity matching based on previously learned information / data. This requires a model where there is a learning phase, before it is actually used. In comparison to the human brain, such a model misses the continuous learning ability, which allows for a continuous change to the outcome of the process. Thirdly, this model also misses the decision capabilities [9], because it is not able to reason based on the information provided. So, in summary, only a few aspects of the human brain have so far been created artificially. Considering that this is an incomplete set and that these are even not yet fully constructed, this means that there is still quite some work to be done before machines will ever be able to take over from us humans. Not to mention that there are still a lot of complexities that would need to be dealt with, like the concious and sub-concious parts of the brain and so on.

However, let's be realistic for a second, as humans we have a genetically / inherently lazy side to ourselves, in the sense that if we don't have to do anything, then why should we, after all enjoying life is nicer than "working", right ? So, if we were able to design something that helps us make the world spin while we are able to relax, would anyone then be complaining ? Probably, the real number of complaints would be limited and for those that still do not like this approach, there would be no need to have your electronic-self taking over, you could still decide to do it all yourself. Looking at the bright side, the world may actually even be better off, since these electronic-self's may have a better retaining of history and e.g. decisions made in the past and their consequences and so would allow for better decisions to be taken without emotions getting involved, so even this worse case scenario still sounds like a win-win situation where at least history would no longer continue to repeat.

So, while this section has hopefully convinced you that intelligence is possible, with quite some work to still be done, before it can ever form a threat to humanity, the next session will look in more detail on how it can be achieved in real life.

3.3 Current Systems for Intelligent Computing

While some of the problems of current computing have been explained in the light of currently existing algorithms, one of the questions is how they can provide intelligent computing. In essence, the outcome of any computing lies in the combination of the algorithm running on a specific underlying platform. The better the algorithm matches this underlying platform, the better the overall outcome / performance. However, it has to be said that most current hardware platforms are aimed at number-crunching, so unless the algorithm can "translate" its intelligence requirements into number-crunching tasks, performance will be sub-optimal.

However, taking the definition of intelligence, it is quite obvious that intelligence does not really come from / through number-crunching. After all, if it would have, then intelligence would already have been achieved. Looking at it from a slightly different angle, number-crunching is actually a human invention based on maths, the language of science. And as with any language, there are certain limitations to the language, and this is certainly not different for maths itself. So, considering that maths has a limited view, and only forms part of the human intellectual capabilities, it seems counter intuitive for it to form the basis towards artificially creating a complete picture of "intelligence". To further support this line of thinking, let us look at some basics of the human brain to better understand why intelligence is more than only number-crunching. The human brain is known to be: massively parallel, running at a very slow clock speed (there is not really a clock present, but it "runs" slowly), capable of intelligence and has very short response times. So, what does that teach us: 1) parallelism is essential; 2) intelligence is possible, and this with little power and short response times; 3) a clock is not really essential, although some "synchronisation" method is essential. More about the lack of a clock later, as this brings along some interesting challenges. Mind you that this is only a snap view of some of the basic concepts of the brain, while a more detailed understanding of the human brain and its intelligence, can be obtained through reading: "On Intelligence" [6].

So, if the brain is capable of achieving intelligence, using a large amount of parallelism and that with a low power requirement, then how does it achieve this ? First of all, one needs to be aware of two of the functionalities of the brain when it comes to retaining information: namely the spatial and temporal locality. Additionally, there are multiple layers which all perform the same tasks, but at different levels of abstraction. For HTM this model is shown in Figure 3.3. The key message here is that according to the HTM model and "On Intelligence", the human brain's neo-cortex would consist of 6 different layers, which is consequently also what is advised for HTM applications. However, each one of these layers would receive certain incoming data, which is compared against the stored information to determine the closest match. This closest match is then passed on to the next level, which takes the output of several nodes from the layer below and in essence works at a higher abstraction level, but performs the same tasks. Each of the levels simply perform these tasks based on incoming data, without having an overview, although the system as a whole provides a way of determining similarity between input and learned data. To be fully equivalent to the human brain, the layered model would differ in the interconnections between the different layers and also allow for two-way traffic, meaning that while information is fed upwards from one layer to the next, there is also information fed back down from the next layer(s) up. Such feedback mechanisms are well known in electronic designs and also here they serve the pur-

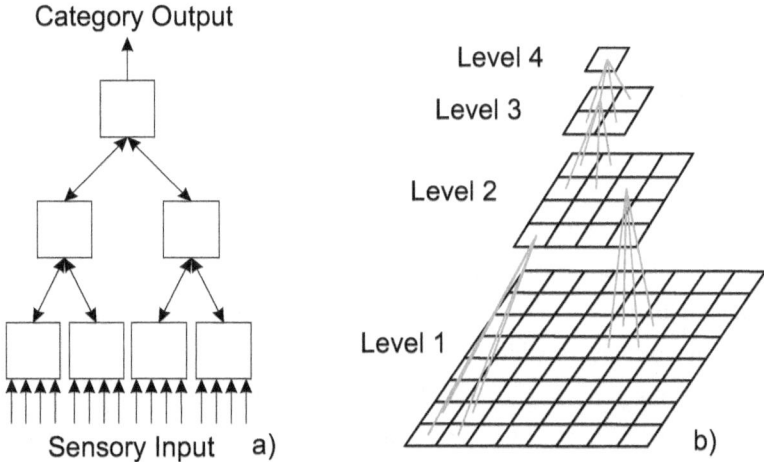

Figure 3.3: Hierarchical Temporal Memory 2-D (a) and 3-D (b) Network

pose of correcting / improving the output of an individual layer. The reason why the output can be improved through feedback lies in the fact that this feedback signal comes from a level that operates at a higher abstraction, which means that it has a better "overview" and can therefore notice if the output of an underlying node is not quite in line with the output of other nodes, at which point corrections can be suggested, leading to a different outcome.

This feedback mechanism is crucial to the proper operation and delivery of such intelligence. It also indicates the importance of connectivity within the (artificial) brain. For example, in a visual application the lowest level detects lines, corners and so on. Now, if the input has an exact match with one of the stored items, then that is great, but that will only happen if the input is exactly the same as one of the stored patterns, which is very unlikely to occur in real life. If one then assumes the case where only a few pixels are missing, then the match would no longer be perfect, but in essence would still lead to the correct result. Correction would namely ensure that the image is auto-completed to its original state, by filling in the missing pixels. The feedback mechanism to ensure this correction can lead to a substantial amount of back and forth communication between higher and lower levels in the hierarchy, which essentially allows for the detection accuracy to be improved. After all, the human brain has many of these correction mechanisms built in, and most of them function without us even realising their existence. For example: when we learn about traffic signs, we will learn all the signs from a book and that will have the signs printed in particular colours and to a certain size. If we then go onto the street, then these signs are likely to be different in size, depending on distance and angle, but are also likely to look different under different light conditions. The easiest test to clearly see these differences is to take a digital camera and take pictures of the same sign in different light conditions (from bright sun all the way to night), and then check what the RGB values are for e.g. the red part of the sign. In each of these cases those values are likely to be slightly different, but to us humans, we tend to "correct" this back to the red that we saw when we learned the sign, since we automatically adjust for different light conditions. Like this, there are many auto-corrections that happen within the human brain and that would need to be implemented if one was to replicate the human brain properly.

When one looks at running HTM in its original form on hardware, then several challenges are faced. Firstly, there is a large set of simultaneous comparisons required between incoming and stored data to determine similarity. This similarity detection can be done in a variety of different manners, but is in essence statistical, and more about that in the next section. Performing these similarity determinations requires a large amount of data to be accessed, and if that was to be performed on a computer with the request-model to request the information / data, to then perform the similarity calculation, store the result and then perform the next comparison and so on ... then there will be a lot of clock cycles required, unless one would manage to massively exploit parallelism. While parallelism could reduce the requirements, it could be challenging to determine the correct amount of parallelism for hardware implementation, since the amount of data to be compared with will be dependent on the outcome of the data stored during learning. Associated with the stored learning data there would also be memory requirements, and preferably this memory would be as close as possible to the actual processing platform, but considering that storage size requirements are difficult to determine, the design of an appropriate platform is not straight forward. Thirdly, one should consider the aspect of communication requirements, which will be large between the memory and processing logic for a single node, and then there will also be essential communication between the output of one node and the input of the next and vice versa (to deal with the feedback). So, essentially that would mean that if there is a hierarchical tree structure, and let's assume a binary hierarchical tree, were to be mapped to hardware, then the more layers, the more challenging it becomes to keep communication local, as shown in Figure 3.4.

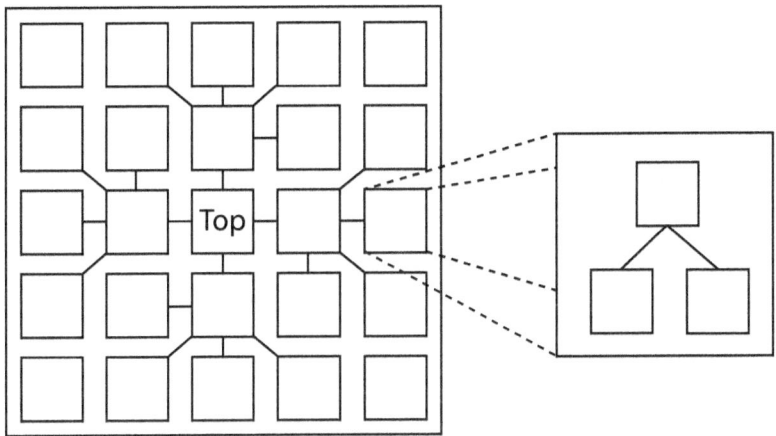

Figure 3.4: Hierarchical Temporal Memory Hardware Implementation Structure for 6-Level Binary Hierarchical Structure, where each Node is Composed of a Set of 3 Nodes, each with Spatial and Temporal Pooler.

So, implementing an algorithm like HTM on a number-crunching platform is particularly challenging, and this lies largely in the mismatch between what one wants to achieve and what the current platforms offer, and then the issue of scalability has still not been approached, since one would obviously want to scale the system up towards larger capacity. From the actual HTM model, there is perfect scalability due to the hierarchical nature, which allows for extra layers to be added to the hierarchy to then combine the output of different networks. However, this scalability does not translate that easily on the hardware side because it is fairly straight forward to understand that for the system to be more intelligent, more data would need to be known / stored, and as a consequence the storage and communication challenges only increase. This obviously raises the question as to whether there are other methods out there that could solve these problems, but more about this in the next chapter.

Bayesian networks and neural networks are both also well-known to provide certain levels of intelligence and are also being implemented on existing platforms. In essence, when one wants to use either of these "algorithms", it is key to define the actual network, and ensure that these networks match the requirements of the task. These network parameters have a significant impact on the processing and data-flow requirements, and overall performance of the networks, which again do not scale that easily since they are application specific. While the processing tasks in a neural network may still be limited, those in a Bayesian network are probability based, and so of a very similar nature to those in HTM.

4

Probability Based Computing

"The scientific imagination always restrains itself within the limits of probability."

– Thomas Huxley

From the previous chapters, it should have become clear that the current computing platforms are really good at number-crunching, at least from a processing perspective, since, if there are large data requirements, then one quickly suffers from the Von Neumann bottleneck, which prevents a further performance increase. Meanwhile, large data bandwidths and / or better positioning of processing and dataflow logic can only marginally improve this situation.

When looking at current systems and their intelligence, the true question is not what intelligence is, or whether it has been achieved so far, but how can it be achieved. Currently, most systems that have shown some signs of intelligence seem to have a link with probability.

This brings along a set of questions: 1) why do probability based systems work better than other systems when it comes to showing intelligence; 2) how can they be improved in the future; 3) can they show more intelligence or do they also have their limitations.

4.1 The Applications

Probability based computing is currently already much more integrated than many people would believe, and that because it is to a large extend hidden, just like in the biological context. One of the best examples, is probably what you would have come across during your most recent web-shopping. Namely, when looking at products, then most web-shops will also offer alternatives, or links to products that other customers who bought this product also considered. These proposals are all based on the concept that similar people would buy similar products, so links of what other people bought could hopefully also help to sell these additional / related products. Other examples of probability based computing can be found in the checking for fraud when you use your credit card online, or the spam filter on your email account, just to name a few.

As one can notice, most of the above applications exploit the fact of statistical relevance to differ between good and bad, and in most cases the outcomes are indeed limited to these two options, however further, intermediate levels may be generated during the process. One of the challenges with making such good / bad decisions certainly lies in determining the threshold between good and bad, and so a full ranking of results could certainly be particularly useful. After all, one has to keep in mind that noise may affect the ranking of results and therefore what would have been expected to be the first result is actually not first, but among the first few. On the other hand, if this incorrect ranking happens consistently, then there may be something wrong with the data processing and / or the information stored within the system, which could / should obviously be further investigated.

In the larger picture, probability based computation is a principle of computation that has a lot of potential, as will become clear throughout this chapter. After all, number-crunching platforms can be used for a variety of different applications, even if one would not directly consider that the underlying platform can deal with the task as such, for example in the case of object-recognition. The purpose of developing this new platform would clearly be to improve efficiency, since there is a variety of situations where items need to be compared and similarity determined. Based on similarity outcomes, items can then be ranked and further "processing" performed.

4.2 How to Achieve Probability Based Computing

To achieve similarity determination, required for probability based computing, there is the need for a certain amount of information to "compare" with and a large set of comparisons to be made, resulting in an output that indicates a similarity match. From this description, one can appreciate that this requires a lot of communication and data movement for a single output. Consequently, the optimal solution to this would be a platform that is probably at the other end of the spectrum from our current data-request, number-crunching machines which use floating point numbers to allow for statistical calculations.

The need for floating point numbers within the calculations in no way eases the actual computational load itself, and while one could consider moving towards the use of fixed point or even integers, the real question then becomes what the effect is of the loss of accuracy in terms of achieving the correct outcome / ranking of items. So, essentially the key output of a probability based method lies in the ranking and the respective values which give an indication of the respective differences between the outcomes.

So, while it may be theoretically and practically possible to run probably based computations on our existing platforms, it is far from the match made in heaven and will therefore never be fast and / or efficient. Since it is never a good idea to give up quickly, maybe the addition of extra data bandwidth and processing power for floating point calculations could help, but then again doing this only moves the problem slightly out of sight, which also means that it would not take long before one catches up with the same problem again. So, rather than continuing to walk along what seems to be a dead-end track, it may be useful to dig a bit deeper and see where the problem actually comes from, and this will require a return back to basics.

The purpose of taking you back to basics is really to show you what needs to be seen. In the development of electronics, analogue electronics was pretty much "first", but then when the benefits of "abstracting values" was discovered, digital really took off. The digital abstraction basically provides for a definition of a clear 0 and 1 area and therefore hides to some extend the underlying precision, but also the problems associated with not having this precision. This resulted in the ability to deal with noise more easily, since there is quite a bit of noise required before the system changes output. Secondly, this abstraction simplified many calculations and operations and allows for easy miniaturisation, which was perfect from a commercial perspective. Now, as much as there are benefits to the abstraction, there are also disadvantages, after all, a full spectrum of voltages is scaled down to a simple 0 or 1 with a grey zone in the middle. This means that to represent larger values, multiple of these 0's and 1's are needed, and based on the binary system, one needs a set of bits to represent a certain value. So, essentially, this abstraction has significantly reduced the information density, since one has moved from a multitude of values on the analogue side to two values in the digital. So, why was analogue not an option for computation ? Well, analogue cannot really store values easily, and computing without storage is a massive challenge.

However, the implications of a reduced information density within digital electronics has quite significant consequences, since it impacts on: processing, memory and communication. After all, a single value requires a lot a bits, and each one of these bits needs to be processed, stored and moved. So, still surprised there is von Neumann bottleneck ? Not really, I guess. Now, if the information density of digital lies at the source of the problem, then why not throw it all over board and do things differently ? Great proposal, and even as a digital electronics person, I can only agree to that, but then what is the solution ? Does analogue offer it ? Well, it does offer better density, but is not noise resistant and cannot be used for computing. And so with neither analogue and / or digital not quite "doing it", it sounds like there will be the need for a mixture of some kind, and / or a new kind of abstraction to say the least. When doing this new development, one should aim to keep in mind that it would be good if things can be kept simple, since that is after all one of the main reasons why digital has become more popular than analogue.

Furthermore, in terms of information density, I can hear some of you thinking, but what about multi-valued logic, that increases information density. After all it combines a few bits "together", which are then processed "together". Yes, this can improve the situation by a factor of e.g. 2, which is a step in the right direction, but our problem is kind of larger than a factor of 2. Not to mention that this processing "together", is much harder than it sounds as it is not easy to design the logic that can deal with this higher density of information, without expanding it back to the uncompressed format. And that would bring us back to square one.

So how can probability based computing form a solution then ? Well, one is ranking outcomes of comparisons, so possibly one can reduce precision, although that may depend on specific circumstances, and while such an approach could more naturally fit analogue electronics, there would be benefit to a certain probability abstraction, but this is where one enters an area of more questions than answers.

From a hardware integration perspective probability based computation would also offer a number of advantages. In an earlier chapter, there was an indication of the problems with silicon manufacture, more specifically the faults on the die, either from before or during the chip manufacturing. Currently, this leads to chip areas being de-activated, or complete chips being thrown away, but wouldn't it be great if one would no longer have to throw away anything, but manage to use it all. Well, this would be possible if faults could be tolerated, and that is only possible if you have a system that can deal with them, which is the case if your computation becomes based on the ranking of items. Current digital circuits are very black and / or white and so as soon as a value is shifted to the other side of the spectrum (0 becoming 1 and vice versa), this has an impact on the overall outcome. With probability based calculations, this spectrum is less about extremes, and therefore quite easily tolerates things being slightly mixed up while even allowing for errors to cancel out one another. After all, if a platform supports randomness to some extend, then adding in more randomness will just make the latter disappear in the former. The remaining question then is whether this can easily be achieved with existing circuitry, and the short answer to that is no, but in one of the next chapters there will be more on how one could get there.

As much as there are challenges to overcome on the "processing side" for probability based computation, when a better information density is achieved, then there are also additional benefits within the system in terms of communication requirements, after all one could even achieve compression through the fact that one only has to provide information that was not previously known and e.g. would affect the ranking / outcome. For example, if someone says that a particular person is ill and that this person had worked in an asbestos factory, then the conclusion / link would immediately be clear to all. However, to a friend of this particular person, one would only need to mention that this person is ill, because he knows where he works. On the other side, to a stranger who just found out this person is ill, one would need to mention that he worked in an asbestos factory. So, as from the sounds of it, this approach may provide simplification.

A system will however only be truly simple if it is easy to use, and that is where the abstraction approach of digital has done miracles. Introducing a new abstraction may deliver exactly the same, with the added benefit that if this abstraction would be slightly more complicated, the more challenging parts could probably still be hidden from the user and only touched by "the limited few" who need to. Simplicity would further be enhanced if the system would be built from a single type of cell.

4.3 How the Functionality will be Delivered

Probability based systems could not only change computing as such, but possibly also programming as it is currently known and used to deliver different computational functionalities. After all, programming is used to translate the "what we want" into "how we can do it with current computers", and this translation largely depends on human interaction. Considering that there are a lot of people employed in this area, putting a stop to that or asking them to re-educate themselves could lead to a massive upheaval, however these programmers generally are used to going along with the latest trend, as long as this change is for the better. So, what changes may be coming up then, as after all, if one was to break with the usual, then maybe it is better to do it properly, since the whole requirement of back-wards compatibility with previous generations of 8086 processors has certainly not helped progress for the better all the time. So, how would probability based, i.e. more intelligent devices deliver functionality towards us humans ?

Well, before going into detail of how to achieve this, it is worth assuming that it would be done through programming, and considering that the person who designs / programmes a device is likely to be smarter than what he / she designs / creates, after all the opposite is quite counter-intuitive. Similarly, one could say if there is a team-effort involved that the outcome would not be more than some kind of combination of the contributions of the team members and so while this can be higher than the contribution of an individual member, it is unlikely to be the exact multiple of the number of members of the team themselves, since: 1) they will be bringing overlapping skills to the team and 2) if they worked on their own they would not create something equally good as themselves, so there is no way they would only achieve something equivalent to the sum of their individual contributions. So, essentially, it will be hard for a team of people to design a system that would "out-smart" themselves, and so essentially we come to the same conclusion saying that it is unlikely for an artificial system to become smarter than us humans and therefore able to rule the world.

When looking at the development of probability based systems to deliver intelligent functionality, the question is how such a system would develop and deliver its "intelligence". To deal with this problem it is probably best to take a look at an example system that is intelligent and from first impressions may actually be probability based too, namely: the human brain. So, while one could argue that there is a certain intelligence built-in, versus an even more significant part that comes from learning. One could then also look at what the concious and sub-concious parts of the brain entail, what is known versus what is learned, and how knowledge is different from common-sense and so on. Now, while these are one by one very interesting topics to look into, at this stage, I would like to focus purely on the aspect of learning.

For us humans, the process of learning requires repetition, which is due to its underlying architecture. After all, neural networks, and then more specifically the network part needs to be created, which is achieved through the fact that neurons have the principle of "What fires together, wires together". This means that by firing these neural connections over and over, they will eventually link up, which then also means that at this stage, the information has been learned / remembered.

In essence, from a memory perspective, as soon as learning takes place in a system, like HTM, then it is important to determine what is repetitive and therefore important information. While it is important to store this information, it is also important to ensure that information is stored only once, to limit the memory requirements. Considering that probability based systems are based around similarity determination, it is probably also best if the data would be stored in a way beneficial to this principle. This is most likely to be found when using a principle similar to that of an associative memory, which also closely represents what the human brain seems to be using. Practically speaking, this means that actually all information is retained and retrieved through association. So, if I would pose you the question: "What were you doing yesterday at 16:13 ?", you are most likely to struggle answering this question, unless it was: "reading this book". On the other hand, if the question was: "What did you do the last time you went to the beach / a restaurant / ... ?" then you can probably start telling me a full story straight away and that with more or less details added to it. This is just one of the ways that indicates our brain is not structured by "addressing" items, but through linking information with other information, creating a huge web of information.

So, while there is the aspect of the actual data storage, a learning system will also need to be provided with training data, and getting this data right is important towards the learning process and the further operation of the system. After all, if we would be told the wrong things in school, it would be what we remember, and maybe we would remember it even better, after all if a kid is learning a language then it always seems to immediately remember the words you don't want it to remember. The good news is that this means we are still in control of what the machine learns / knows, and consequently will be capable of "doing". However, it will probably not be economically feasible to put machines into schools to learn, so another mechanism may need to be found. Now, storing the learned information is certainly not that straight-forward especially when one considers there is the need for a certain flexibility. After all, if the machine is to recognise dogs, it would be impossible to show it all different angles of all different dogs on this planet during the learning process. So, it will be key to determine what is essential to the learning and then also include a certain level of flexibility within the system so that it can detect based on "concept" rather than pure data. And this is also partly where the feedback mechanism, as discussed previously, comes in. After all the colour correction of traffic signs, is one of those mechanisms that allows for such flexibility. Similarly, the human brain automatically scales objects, so if the object is learned at a certain size, then it can be recognised even when presented at a smaller / larger size, e.g. most of us have seen the Eiffel tower on a picture before we saw it for real (assume you have) and we still recognise it (shame on you if you didn't). Another correction lies in "automatically" rotating objects to recognise them, e.g. if a chair stands upside down, we still recognise it as a chair. As you hopefully start to appreciate there is a huge set of these "correction" mechanisms around and they seem to help in reducing the amount of information that needs to be stored.

The question then becomes as to how and when these automated corrections are / need to be performed. The initial belief would probably be that these corrections need to be performed as pre-processing steps, before an actual recognition takes, which then raises the question as to how that can be achieved and what type of systems would be required to do this. However, considering that one only knows the information available when it has been processed, which in this case refers more to the recognition aspect, how would one know what to correct if one does not even know what it is ? This may indicate that these corrections are actually to be built into the system that has learned and recognises, which is also what HTM seems to indicate. Although, further research will need to show the best approach.

Now from the actual aspect of storing the data, this flexibility requirement certainly raises concerns as to what should then be the standard for storage. After all, following this principle requirement, it is not possible to store e.g. pure picture information, as that would for example lead to problems in terms of scalability. On the other side of the spectrum would probably be a definition of the object, but this raises two problems: 1) how can one check against a definition; and 2) how does one create such a definition in the first place. Illustrating this point with an example: a table. Most of us would have grown up seeing a table of some shape or form and our parents would have told us that that was a table, but then you probably found yourself coming across another table, with more / less legs (standard 1 to 4 (or more)) and you still recognised it as a table. So, what is then the definition of a table: An item with a surface on which we can put items (to eat) and which has at least 1 leg ? Well, then how about a table that is suspended from the floor, is cantilevered from another object, and / or a desk or kitchen surface which has food on it ? So, one could decide that the definition needs more detail, like: a flat surface with a space underneath that is held up normally through the use of legs, and is used to eat and socialise around. Now, making this definition more detailed also makes things more complicated, but one can still find counter examples, like a desk. Get the point ? Indeed definitions will probably not do very well. So, there certainly lies a challenge in finding mechanisms to store the data.

Having spoken about the need for learning, the importance of choosing the appropriate data for the learning process and some of the actual storage requirements, it is probably a good idea to also consider when learning should happen. One could choose to have learning happening in a controlled environment, yes, like schools, but not quite the same. A controlled environment, in this case, refers to a controlled time during which learning takes place, which generally means that the machine would learn at the very start, so before it becomes operational, and then following that, one could decide to have controlled interruptions to teach it more. The other, and preferred option is that following an initial period of learning, the machine continues to learn as it goes along, although that does bring along additional challenges.

Coming back to the topic of what is "pre-programmed" in the human brain and what is added later on, one can start to wonder what type of tasks a machine would be able to learn and whether that would be anywhere close to what we humans can. Well, if the machine is well designed, then it can probably achieve a large set of tasks, even calculus, translation and so on. That would then also mean that these new machines compete with the existing number crunching machines. However, before even competing, it would be essential to understand how intelligent machines may be able to perform calculations. One way to determine this is by understand how the brain of those "exceptional humans" that can give you the square root of: 4 325 298 289 in less than a second, works. Some of these people claim they get a visual representation of the number and that is how it "works" in their brain, and maybe that is the way to go, until then, I had to use my calculator, and to make sure you can continue reading, the answer is: 65 767.

5

Relevant Computing Developments

"Why compare yourself with others? No one in the entire world can do a better job of being you than you."

– Unknown

The previous chapters gave an overview of current systems and introduced the concept of machine intelligence and how this can possibly be achieved through using probability based computing. This chapter will provide a bit more detail on some fields already touched upon that have often been researched for a while, but not always quite "delivered" yet. After all, it seems there are a considerable amount of people who believe that by continuously trying to improve the same old, one will eventually manage to crack the nut and achieve much more. Before one should even try to continue this way, the big question is whether there is still a nut to be cracked and / or how much more of the same can be done before someone will realise that this is just not the way to continue.

Before going into any further details, please note that the overview given in this chapter is by no means meant to be complete or cover the approaches in detail, but will focus on the points that are relevant in the context of this book.

5.1 Bayesian Networks

Bayesian Networks (BN) are a very well-known method of probabilistic computing. They have been around for quite a while and are used in a variety of different contexts. When comparing Bayesian Networks with e.g. HTM, it appears that HTM does not only require less human input during the design phase, but also leads to better recognition results [13]. The popularity of BN however lies in the clear indication of it being a probabilistic based model and this has even led to some hardware being constructed to specifically provide this functionality [19].

The probabilistic model of Bayesian Networks lies in the probabilistic directed graph. So, while the graph connections indicate a relation between different items, each connection also has a value associated with it and this value indicates the likelihood / probability that the connection path will to be followed. So, considering that these paths connect different states / events, the probability value indicates how likely it is that another event occurs when the current state / event has been reached. Consequently, certain paths can be created through this graph, all with a certain probability. During the learning or training phase, it is these probability values that will be customised to ensure that the BN aligns more closely with reality.

In essence, there are multiple levels of design possible for using Bayesian Networks. If the task is not to complicated, and could be achieved through a small network, then on average this network would be designed by a human who would draw the graph and if possible also provide the probability based data for each of the connections. When the network increases, then the amount of automation also increases, and consequently allows for probability values to be learned through a machine learning process and when the network becomes even more complicated then also the construction of the graph would be automated. So, depending on the system / application to be designed, the resulting graph can be completely different, which also makes the actual implementation in hardware / software to be particularly challenging due to the network communication having to change dramatically from one network to another. Current hardware has already shown that such flexible communication requirements result in a lower performance, especially if a larger system is being mapped onto a device. If the device implementation is limited in size, then this would require more flexibility in swapping structures in and out, which brings us again very close to the current computing system. So, while a dedicated hardware development for BN may reduce the number of transistors required to perform the actual computation, the model itself seems to suffer from some of the same problems that the Von Neumann model struggles with. Hence, performance increases would be limited to the reduced number of transistors used in a dedicated HW implementation and the possibility of requiring smaller network sizes.

5.2 Cellular Automata

One could argue that to achieve probability, a certain "randomness" is required, which could be achieved in various ways, for example through the use of the randomness in CMOS transistors [16]. However, intuitively, most people would believe that good randomness comes from complicated structures, which is actually far from true when one looks at e.g. cellular automata.

Cellular Automata (CA) has known several ups and downs in popularity over the years. The last up is probably related to the publication of: "A New Kind of Science" [23]. As much as the book forms an interesting read for quite a few days, one should be aware that this was not as new as it may sound. After all, people like Alan Turing and Von Neumann were already looking at Cellular Automata [20], and their publications can certainly bring an interesting perspective to the field, not to mention that one could actually consider our current digital computers to be special cases of cellular automata machines.

The statement of what's in a name also applies to cellular automata. Firstly, it is cellular, so there are "cells" involved and secondly, there is an aspect of automation involved. In practice this nails down to the fact that there are a set of simple rules, as shown in the top part of Figure 5.1. Starting from a certain start condition, the rule-set is checked for each set of cells, which leads to a second line of information, which can then be used as an input for the same set of rules and so on. So, eventually, depending on the rule-set, a certain pattern evolves. This pattern can be very repetitive and / or seem completely random, which proofs the fact that complex patterns do not require complex structures to generate them. The rules themselves can be very simple, with only black or white cells, which essentially corresponds with boolean logic, over rules that cover 3 cells, to rules that cover more than 3 cells and multiple grey-scale levels, more closely similar to multi-valued logic. When more extensive rules are used, then the patterns more quickly become more extensive. However, it is important to note that the rule sets remain simple and can therefore be easily executed in parallel and generate very visually attractive patterns. However, the full applicability of CA towards more efficient computing is still unknown till date.

In a computational context, "computing" implies that for a given input, there will be a certain output, which matches expectations, based on the "algorithm" being performed. For example, when an addition is performed, then the output is the sum of the two inputs. For binary computation, this process is fairly simple, since it manipulates binary data, using local "bits" and an outcome is generated according to a simple rule. Hence, one could argue that current digital computers are a very special case of cellular automata, but then with very severe boundaries in terms of the rule-set and number of iterations. Cellular automata are however capable of more, especially since there is a much larger set of rules available and these rules can all run continuously. However, the link between this rule set and the actual "computation" performed is not yet clear from a human perspective. In other words, if there is a requirement for a specific task, then it is not clear as to what rule set is required and how many iterations of this rule set are needed, which could be an issue of not having sufficiently understood the "language" used within cellular automata. This lack of a clear link

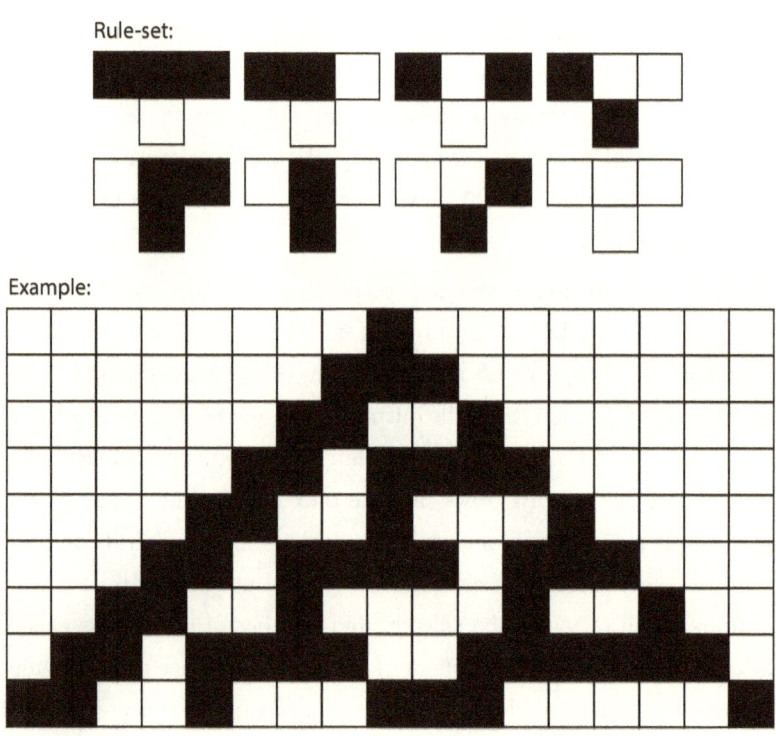

Figure 5.1: Cellular Automata Rule Set

between rule set and outcome also determines the "time to compute", which currently corresponds to "performance", and has been one of the main drivers towards improvement over recent years. Current systems all perform their functions within a certain amount of time, whereas in the context of cellular automata the time until a result is achieved is one of the unknowns. While the concept of time within a system should be of no particular relevance, the important part lies in the fact that seen from externally, the response time is acceptable. However, the concept of time in current computing systems is probably quite different from what it is in the human brain [1]. After all, current digital circuits have very fast clocks which are used to trigger the design in "finishing" and starting its calculation due to the introduction of pipe-lining, and while this eases the design quite extensively, it also brings along additional challenges.

A second key aspect of what Stephen Wolfram mentions in his book on cellular automata lies in the fact that cellular automata form the basis of all biological development. The theory builds on the ability of using cellular automata as chemical and therefore analogue cells. The explanation goes as follows: if one looks at the start of a living being, then there is in essence the combination of 2 cells, which eventually leads to a large biological system, with many different organs, and functionalities. The big question then obviously is how one goes from a few cells to many cells with different functionalities. In the most basic of answers, there is cell division, but then how does a cell know which organ it will be part of ? While the DNA itself is simply "copied" during cell division, each cell contains a set of chromosomes which in turn comprise genes. These genes determine the function of a cell through the way it expresses or "manipulates" proteins. Regulatory regions in the DNA act like switches that turn a gene expression up or down. The state of these switches then also influences the role of the cell, but it remains largely unclear as to how these switches are "controlled". In conclusion it is a set of simple rules that are being repeated, which lie at the basis of controlling the cell functionality in this large duplication / division process. Consequently, large, complex biological beings, like humans can be "created". Wolfram goes even further in his claims, when he mentions that the Cellular Automata principle lies at the basis of every biological system and so if biology is to form any inspiration to improve computing, then this field can certainly not be ignored.

So, as much as cellular automata may provide some useful tools, the true question lies in how to use this tool-kit. Some of this may depend on getting a better understanding of the "language" of cellular automata, but in essence cellular automata are not that different from finite automata, to many better known as finite state machines. Essentially, these are state machines that provide a certain relationship between input and output, based on simple rules. While each of the states are important as recalling the particular status of the system, the transitions as such are generally controlled by very simple rules, which normally depend on the input to the system. These states as well as transitions then determine the output of the system. Currently, the main use for finite state machines lies in their use to control circuits where they ensure that algorithms follow a correct sequence. So, how would finite state machines behave as a computing model ? Well for one, considering they are finite takes one worry away in comparison to the cellular automata principle. While they have memory included and allow one to retract his path through the state machine in a "feedback" manner, most of the finite state machines do not really deal with parallelism, while their data input and output requirements are on average also quite limited.

5.3 Quantum Computing

For quite a long time, it has been said that Quantum Computing is the "next" generation of computing, but progress towards this goal has unfortunately been rather limited. A quantum computer is a device that directly exploits quantum-mechanical phenomena, such as superposition and entanglement. In practice, not many quantum computers have been produced to date, and that is largely due to the fact that they generally operate best at a temperature close to 0 Kelvin (way below 0°C), because at this temperature the thermodynamic influences are limited and these are the most optimal conditions for quantum coherence. These temperatures may be obtainable within certain lab environments, but are certainly not practical in a large set of contexts, let alone for a consumer device. A second limitation lies in the fact that the number of calculations that can be performed in sequence is limited by the level of quantum coherence that can be achieved, where quantum coherence refers to the quantum remaining constant over time (obviously while it is not used for calculation). The latter does not form to much of an issue yet, since considering that the technology is still in its early days, one wants to check the result of each individual computation to ensure correct operation. However, there is a second aspect that is much more important in relation to the future of quantum computing and this lies in the so-called "observer effect".

In short, the observer effect means that as soon as one measures something, then it will have been affected by the measurement. For example, if one tries to measure the pressure of a tyre, then that would not be possible without releasing some air, which consequently changes the pressure. The same applies in electronics, since measuring voltage / current affects the circuit, because there is not a single measuring device that has infinite or zero resistance for voltage / current measurements respectively. The same applies in quantum computing, where generally the spin (either left or right) of a quanta (minimum amount of any physical entity involved in an interaction) will contain the "essential" information. When several quanta interact, their spin can change, which eventually is the way that computing is performed in this context. However, measuring the spin, affects it, and therefore prevents the calculation from being continued.

The observer effect also links up with the fact that a quantum state of the system actually is a superposition of all the available states, which in essence means that a physical system, such as an electron, exists in all of its possible theoretical states simultaneously, but when measured or observed, it only gives a result corresponding to one of these possible configurations. As complex as this may sound, it means that in essence an electron is always in multiple states simultaneously, while one of them is more prominent, which is the one being "observed" / measured. The latter implies that the "observed" state has a stronger "focus" or is "ranked higher" and this actually links quite nicely with the requirements of probability based computing [18]. So, it seems that even quantum computing is inherently ready to be used for probability based computing, which is of no surprise considering that probability lies at the source of most things in nature and the quantum effects therefore form no difference to this.

However, there are more aspects that make quantum physics / computing interesting. First of all, quantum physics is one of the few theorems that manages to explain pretty much everything on our planet, which cannot quite be said from "old / existing" physics. One example of such an interesting aspect is entanglement. This basically refers to the fact that two physical systems, can be "linked" without being physically connected, or in close proximity to one another. However, if these items are entangled, then that means they are "indirectly linked" with one another, which means that if one changes state the other will also change state to be in an identical state to the one it is linked with. This "linking", which is essential for the entanglement can only be achieved when both particles emerge from the same event, one could say that this is very similar to one of as set of twins feeling what the other one feels. However, as a concept this brings along interesting perspectives, especially at the front of communication, since the "alignment" of these two particles through entanglements allows communication which is faster than the speed of light, and could possibly be expanded to combine more than 2 particles. In essence this could revolutionise communication, since there is no longer a need to physically send bits of information from source to destination and vice versa, because both would continuously stay in sync with one another. This could certainly overcome many of the communication problems that computing currently faces and that are unlikely to be solved until communication with infinite bandwidth and speed, like "entanglement" comes along.

Indeed, quantum computing can bring us quite a few interesting aspects, and hopefully the issue of needing excessive cooling can be resolved rather soon. Even if it is still to take a while, the concepts of probability based computing seem inherent in most natural systems and that includes quantum computing. Therefore, intelligent, probability based platforms can be considered a continuum for many years to come.

5.4 Artificial Intelligence

Artificial Intelligence is more of a concept that comprises a lot of different research fields, although this is not always the way it is acknowledged or seen. So, while a lot can be classified under this umbrella, not necessarily all the work that would and should be fitted here, necessarily wants to be associated with the AI heading for various reasons.

So, just to give a few examples of what does fall under this naming: autonomous / distributed systems, machine vision, robotics, as well as pretty much anything with intelligent / intelligence in the name, like: intelligent agents, and so on. Please be aware that a lot of what is called "intelligent" in the mentioned examples does not necessarily comply with the wider definition of intelligence used within this book. Just to give an example: when looking at autonomous / distributed systems, then this implies that one moves away from a centralised unit that controls every single action within the system. The main benefit of this approach lies in the fact that if each of the units within a system becomes more autonomous then the bottleneck associated with a centralised control is overcome. However, taking that each of the units now operates independently, then there is the need for a mechanism to communicate and / or synchronise the operations across the different nodes. In this context, a network proves to be more beneficial, and while this may seem like a natural extension to our autonomous system, this is also what is sometimes referred to as: intelligent agents. Where the agent is the distributed controller, and the intelligence lies in the fact that they are autonomous, while still communicating with other agents to optimise their own behaviour.

Most of the developments presented in this book obviously fall under the umbrella of "Artificial Intelligence" for the fact that they are all artificial and relate to intelligence. The only issue to keep in mind is what definition of "intelligence" is being used in the specific context.

5.5 Neural Networks

Similar to AI, neural networks have been around for quite a while, which is in itself interesting. Personally, I think one of the biggest challenges is to build a proper Neural Network, considering that there is still so little known about the actual brain and its functioning at the lowest level. However, quite a few research projects are looking into mapping as many as possible neurons onto silicon and ensuring they have good interconnectivity [15]. These projects are meant to learn more about neural networks through building them, hoping to eventually build an artificial brain. Any of these models duplicate a system in which the neurons only link if they have been fired multiple times to "wire" together, which allows for a training phase and an actual using phase. While neural networks are not quite common practice yet, they are used in specific settings where a learning based approach is most appropriate. While originally they were implemented on platforms, such as FPGAs, more and more efforts are put into constructing them directly onto silicon [22].

At the time of writing, there are several large research projects, around the world, that focus on better understanding the human brain and building an artificial version of it. The scale of these projects is large, and for example in the US, several projects are working on "constructing" the brain of a mouse or cat and it is expected that this would require a massive room filled with high-performance computers to enable modelling this brain at the neuron level. While there may be benefits to performing this, the artificial system certainly seems to be significantly larger in comparison to what nature manages to achieve. Firstly, the large set of hardware resources will require massive amounts of energy, which stands in sharp contrast with the energy efficiency of the brain itself. It is probably not difficult to understand that modelling is always particularly more challenging than doing something that exactly matches the underlying platform. However, a few of the challenges have already been mentioned. For example, essentially, one tries to model a 3-D structure on what is pretty much a 2-D platform, which certainly brings along communication challenges, and to resolve these, relatively long distance, high power communications would be required in abundance. Secondly, there is the aspect of information density, which is probably quite a bit lower in a computer context than it is in the real brain, therefore this again would lead to a significant scaling up for the model in comparison to the real thing. Not to mention that considering the brain is a communication centric device, and current computers are processor centric, this mapping will be particularly challenging. Each of these will obviously play a part towards larger power consumption for the artificial equivalent. Furthermore, projects of this size, also raise the question as to which part of the brain is actually being implemented, since the brain has many different areas, dealing with different functions. Normally, the frontal cortex is considered as the more interesting part, but the question is whether this can provide functionality on its own, and if so what that functionality would be. So, while building a better understanding of the human brain itself is essential, there is certainly still quite a way to go before a fully artificial model will be constructed that is as energy efficient and as small as an actual brain.

5.6 Reservoir Computing

Reservoir computing is not as popular a name as for example neural networks, but essentially reservoir computing is a specific case of a neural network. While the key to "reservoir computing" lies in the reservoir, which is a neural network that will be trained accordingly. After the neural network, there is a read-out layer, as shown in Figure 5.2, which ensures that the output becomes understandable to us humans. Now, while the links within the reservoir itself are actually fixed, only the links between the reservoir and the readout layer are being trained. This allows for a fixed structure within the reservoir itself, meanwhile reducing the overall amount of training. Drawing the analogy with

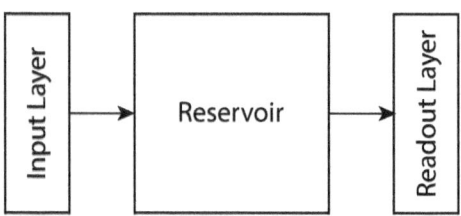

Figure 5.2: Reservoir Computing Structure

current architectures, and especially the fact that each structure has its specific strengths, it would be strange if this does not apply in this context. This would probably mean that this "limited" flexibility will allow for certain types of computing to be performed well, while others will simply require a different type of reservoir network. More research is still required as to how flexible this fixed network would be in practice.

5.7 Reversible Computing

Reversible computing focuses particularly on the previously mentioned aspect of feedback, since reversible computing ensures that there is the ability to reverse the computation and go from output back to original input, which stands in contrast to most currently available systems, which are not reversible. For example, when a current processor achieves the outcome of 5 for a specific addition, then there is no indication as to whether this came from $4+1$, $1+4$, $2+3$, etc. Besides the fact that this naturally ties in well with the feedback principle, there is a second reason why reversible computing provides benefits, and this lies in the physical aspect of retaining energy.

Current irreversible computers produce a lot of heat during their computation and this lies to a large extend in the irreversibility of their computations. This generated heat then leads to problems about cooling, but also means that the silicon degrades much quicker since it cannot stand being at high temperatures for long periods of time. Eventually, this means that the circuit may no longer perform what it was designed to do. So, the goal is to keep the silicon as cool as possible, but that means that extra energy is required to cool the device, which easily doubles the originally intended energy requirement. Hence, it would be most beneficial if the amount of heat produced during computation could be minimised.

Reversible computing allows for a minimisation of produced heat, since it is adiabatic, which means that the process occurs without the transfer of heat or matter. Now, while perfect reversibility may not be easy to reach, one can intuitively understand that if there are a certain number of inputs converged into an output, then there will be a certain amount of energy in excess. If these same inputs are converted into different outputs, namely one that is used for the next computation and one that can be used to revert the computation, then the energy levels have to stay the same, since one has to be able to revert back to the place of start, which also means the same energy levels as available at the start, so one cannot waste energy, or it would become impossible to return. So, to solve the heat problem with current computers, one needs to ensure that computation moves energy around, rather than creating "waste". The question then is whether this is a model that can be used for computing. To which the short answer is: yes, because of the fact that this reversibility is no different from the feedback mechanism previously discussed, which is essential towards improving the outcome of computations in a brain-like / intelligent setting. The principle of retaining the state of energy within a system can also be found back in quantum computing where there is the necessity for coherence, meaning that energy is moved between different locations to ensure this coherence, and consequently there are no losses, which generalises the concept of reversible computing towards adiabatic computing.

6

The Future

"The true sign of intelligence is not knowledge but imagination"

– A. Einstein

Having looked at current architectures, and their problems as well as why probability based architectures form a good future development, the next question lies in how this can / will be achieved, and more specifically what will need to be taken care of along the way. Through the previous chapters, several concepts have been repeated in various contexts and achieving these will provide for different challenges.

From a computing perspective, probability based approaches are much more prone to errors, noise and all other "malware" that has been avoided in the current black / white systems for the reasons previously explained. After all, the world is simply not perfect, and maybe it is about time that also computers learned to live / deal with these imperfections. While looking at a few conceptual ideas which may (not) need addressing, it is quite certain that there is a need for change, which will require new ways of thinking.

6.1 Designing with / for Imperfections

When looking at current designs, then they are quite "easy" to design, after all the basic principles are very much black / white, which even translates in them either working or not working. While this may have influenced some of the "decisions" to go down this very route at the start, based on the information provided in the earlier chapters, it should be clear that this path is probably not going to continue to work, hence there is a need for change. So, in order to contrast with a black / white model one would then need a "greyish" type model, like a probability based one, since it is more likely to deal well with "less certain" circumstances.

One could consider that the designing of a system able to deal with imperfections is going to require much more complicated designs, since "extra" circuitry would be required to deal with these imperfections. This would indeed be the case if such circuits are required to "filter" out the imperfections and bring it all back to the black / white model that is currently used. However, if the underlying model is more error prone, because it is essentially based on errors, in the largest sense of the word, then that would be more of a well found match. After all, the probability model and the actual learning that takes place within the system, provides the system with several layers of abstraction, which would hide the underlying imperfections from a user's perspective. However, not all these imperfections may be present at the learning / design phase, and so it may be essential for a system to re-learn, or adapt its behaviour, based on more recent circumstances, which would certainly not form a problem, but potentially only a challenge in finding a way to allow continuous learning. Ultimately, the probability based principle will rank results, and such ranking does not necessarily provide for correct absolute values, it requires the ranking to be approximately correct. One can get away with this approximation because feedback can provide for corrections to the ranking itself and the eventual outcome, since the system operates on the principle of likelihood rather than exact outcome.

So, in essence, to most, a probability based platform could be treated like a black box. Only if one was to design the black box, then one would need to consider the essentials of what goes in, and even this should not be a very complex structure, but a simple cell that is repeated so that the larger system can deal with all input / output requirements. Additionally, once a basic cell has been designed and shown to work on small examples, then this principle can easily be replicated, and even expanded upon with additional devices to deal with the requirements of the specific design set-up. After all, in the ideal case, there is no need for a separation between memory, communication and processing, and so an architecture based on a simple cell, would also make for the perfect reconfigurable device.

6.2 Information Density

As mentioned in some of the previous chapters, information density plays a significant part in all aspects of computing. Since a higher density can lead to "higher" communication bandwidth, more effective storage and processing. When comparing current digital / binary systems with the human brain, then it nearly seems as if those are at two extremes, while one stores a single piece of boolean information and needs to replicate that principle many times over to deal with the more common type of information used in systems, the brain seems to have multiple dimensions in which information is represented, namely: chemical, electrical and so on. In improving our current systems, it will therefore be key to ensure a more condensed information density.

To many, the natural reaction would be to use analogue electronics, however, this approach would suffer too much in noisy conditions. Considering that a probability based system is in itself based on an "error principle", the question then is how much noise would affect the output. In the best case, there would be no significant impact, while in the worse case one would probably fall back to the current situation where the system is likely to continue operating fine until a certain threshold has been surpassed and problems will start to appear.

The actual information density problem is more of a lower level problem, since it affects how things should be implemented in actual hardware. Therefore, it is unlikely to affect the actual users in any negative manner, as the past has shown that an abstraction, like used in digital electronics, is particularly beneficial. This is also what probability based computing offers, but then with the abstraction having a slightly different definition. If resolved well, then the increase of information density could also bring benefits in terms of power consumption, performance and so on, because essentially the amount of components required for information processing could be lower. This would also mean this part of computing is more in line with nature and therefore works more efficiently, while also showing more "intelligence" than current systems do.

6.3 Feedback

While many electronic systems only work well when using feedback, as a concept, it is not well integrated into computing, even though the benefits could be numerous. Feedback also has a slightly different meaning in electronics from its meaning in biological systems, where it refers to the process of moving information back and forth between different information processing layers. The latter definition aligns better with the concept of reversibility in the computing context, so for the purpose of this discussion the concept of feedback is used in its biological meaning for the computing context. The first benefit of feedback is fully in line with its benefits in other contexts, namely the improved precision / outcome. In a probabilistic context, if higher layers can feed back information to lower layers, based on the fact that they have a better overview of the situation, then noise can more easily be filtered out. To achieve this, one does not only need to be able to calculate back-wards (from output back to input), but also check the appropriateness of these new input values versus the original ones. Currently, computing is one way traffic only, going from input to output, and that applies to the actual hardware as well as the sequential execution of instructions on a processor, which in essence are run through in a single sequence, without the option to reverse them.

Including feedback within computing under the form of reversibility, would also make the computing model more adiabatic. The main benefit of this lies in the fact that adiabatic computing no longer suffers from heat production, since there is no longer a conversion of input energy towards output energy with the excess turned into heat, there is a conversion of energies from multiple inputs towards multiple outputs, retaining the same physical amount of energy all the way through the process. While it may be challenging to achieve full adiabatic computing, it is certainly possible to move much closer towards this model, which would also result in significant energy savings.

Reversibility can be achieved using a variety of different models, based on a different type of underlying cell structure. The most suitable model will however depend on some of the decisions taken in relation to the other key points about the new computer design.

6.4 Timing

As much as timing may not have been explicitly mentioned yet, it is an important aspect in designing a new computing system. The reasons for this lies in the fact that in a clocked system, the clock provides for specific points in time at / by which events should happen and therefore gives structure to the design and the user. After all, the clock, allows for very clear boundaries to the user, since it provides a certain time-frame to the design. On the other hand, this time-frame can lead to limitations in terms of flexibility. For example, the clock speed is generally only changeable within certain limits, which means that speeding up as well as slowing down can be challenging, while this could be particularly beneficial in certain settings. The clock dynamics generally also lead to a substantially larger power consumption, due to the clock related switching of the logic components being reduced. Nevertheless, timing cannot be fully eliminated from any design, but one could wonder whether a move from absolute to relative timing would make sense. In this context, relative timing would refer to the mechanism that deals with what has "completed processing" and what has not. This then stands in contrast to the current mechanism of clocked circuits that run on absolute time, since they follow "a central clock". In technical contexts, this should be associated with synchronous and asynchronous designs respectively.

When introducing relative timing, then the challenge lies in determining the relative indication of when certain calculations / communications are finished. Therefore special protocols would be essential to allow the two parts to inform one another that all is finished and the next steps can be taken. The same would apply to the different stages in processing. While this timing model can sound particularly challenging in comparison to an absolute timing model, the relative timing model fits much better with the requirement towards information density and feedback. After all, the new model for information density may prove that there is a variety of different densities that are required, and depending on the density "processing, communication, ..." can be performed faster / slower. Similarly, when it comes to feedback, the process of passing information back and forth between different stages could be limited, although that is unlikely to lead to a good outcome, because to allow the system to converge to a solution, more or less iterations may be required. Depending on the number of iterations, a different amount of time would need to be spend, and therefore a more flexible, relative timing model would be more suitable.

The remaining question then is how does one manage to design circuits without a clock, without compromising on the ease of design. While the answer to this may not be that straight forward, one should probably consider that pure digital, as well as pure analogue electronics are no longer the future. Hence, the future probably lies in a combination or mixture of both, and while analogue circuitry has till now managed very well to operate without clocks, so do many biological systems. After all, there is still no proof that the human biological clock has anything to do with the operations within the brain or that it would run with a 10GHz clock speed and / or a counter to determine what seconds, minutes, hours and so on mean to our biological system. Consequently, there are still quite a few interesting hours to be spend on this issue, question is whether you are up for this challenge or simply prefer one of the other ones already mentioned or still coming.

6.5 Evolution Algorithms

In the biological setting many years of evolution has led to the "systems" as they currently are. During these years of evolution, biological selection has ensured that the fittest / best systems survive and evolve further, while the not-so-good options do not. While this has been going on for decades, centuries, millennia, it is still ongoing and unlikely to change anywhere soon. After all, this is the principle that has allowed species to adapt to their environment and continue to live / survive.

Consequently, one could ask a similar question as to whether biologically inspired computing systems would not benefit from this type of development, and maybe they would. After all, evolution algorithms do exist and perform exactly this function. They have certain degrees of freedom, i.e. parameters to the system being developed, and a mechanism to determine the "best creations". Eventually the algorithm runs through creating and combining a large set of systems, creating new systems, while each one of the created systems is then tested against certain criteria and ranked as to determine the best solutions. The best solutions are then again combined to improve and so on and so forth. After a large number of iterations the new, most "optimal system" is taken as the output of this development. In a human context this would certainly raise ethical questions, and one could pose the same in this context, but let's assume that for technological development this would not apply. What would be required though is being able to fit the required developments in this context and then let the algorithm run through. While determining these parameters and measuring data could be a challenge, this approach would certainly allow for a larger set of combinations to be checked in comparison to what a human can achieve in the same amount of time. So, if the criteria and parameters are well selected, then this approach could certainly provide the ability to design better systems.

6.6 How to Achieve the Change

Maybe not the type of section you expect in a book, but nevertheless a very important point that many people overlook.

To many people, change, means, what I call: "shuffling their feet". After all, there is enough evidence that science has not progressed as much over the last decades as it used to do before then. While there is a vast array of reasons for that, this does not mean that any of these are appropriate to not change things for the better. After all, and I have said this already, quite different thinking will be needed to overcome the problems that are faced now. For one, this means that if there is a problem that needs solving, then let's solve it by not simply looking at the problem itself, but the actual cause / origin of the problem. Dealing with the origin of the problem, is likely to be harder, but is also more likely to solve it for a longer period ! For example, in relation to the heat being generated in current computers, one could decide to simply add in cooling. However, essentially, there is a physical reason why this heat is produced, so the true question lies in the fact whether the physical problem can be dealt with. Since, tackling the heat as heat will mean that one continuously fights with the need for better cooling systems due to more heat being produced, until one gets to the stage that it becomes a physical limitation, and then that would also limit performance ... and maybe this is exactly the position the world is in now.

On the other hand, you will come across people saying: "these are interesting times", which is based on the fact that there are a significant amount of problems being identified which will need new ideas to solve / overcome them. However, while this seems to identify the need for, there is still a lack for substantial change to break with the old ways of thinking, to allow the existing problems to be overcome. Indeed, step change is not always easy, and there are again specific reason for that. One of them is the fact that step change means there is a need for some certainty about what is being proposed, and that the proposed solution is better than what is currently around, but then this overlooks the fact that failure teaches us lots, if not more, than success, and that especially in a research context. Secondly, change is one of the certainties of life, but is not always dealt with that easily, which is also reflected in a large number of settings, e.g. industry is quite often happy to hold on to existing, proven methods. This generally gets worse if the company's size increases, due to the lack of flexibility in larger structures. So, breaking the current cycle will not be easy, especially considering that the ever increasing pressures to perform, generally captured in numbers, provide for less opportunity to "break away from the usual". After all, breaking away takes more time, money and human energy, because one can no longer be on auto-pilot, but needs to truly invest, but this is not the time or place to get too philosophical. Just remember that only action causes reaction, eventually leading to results, and then you know what to do next, once you finished reading this book ... obviously.

7
Conclusion

"A conclusion is the place where you get tired of thinking."
– A. Bloch

"In literature and in life we ultimately pursue, not conclusions, but beginnings."
– S. Tanenhaus

"People mistakenly assume that their thinking is done by their head; it is actually done by the heart which first dictates the conclusion, then commands the head to provide the reasoning that will defend it"
– A. de Mello

Computing as known to us, and at the time of writing this book is in a bit of a crisis, although very few would acknowledge it. Progress has been lame and most would blame the Von Neumann bottleneck for it. While the true reason for this bottleneck lies in the unequal development of memory and processing, this is actually only a small part of the full picture of problems. Namely, the separation of communication, memory and processing into three separate items is not particularly helping overall development for a variety of reasons, while actually there is not even a need for this separation. Secondly, the current digital computing model has based itself completely on a black / white abstraction, and this abstraction starts to suffer more and more from the fact that it is difficult to keep things pure black or white. This is related to a combination of shrinking feature sizes and other impacts, such as silicon manufacturing problems which all push signals more into a grey region for a variety of reasons, due to noise being added to low operating voltages, manufacturing defects and so on. However, shrinking feature sizes are one of the few ways forward in terms of improvements to electronics, and their performance. Hence, new solutions will need to be found, which turn these greyish, error-prone platforms back into useful devices, which even manage to do more than previously possible.

The current number-crunching platforms have been proven to be particularly good at certain tasks, but the expectations are also here increasing and more intelligence is expected, which means that these platforms should be able to do more human like tasks, while also lowering their energy requirements. Currently, any "problem" or task that should be executed, has to be "translated" to fit the number-crunching model. While this can be done for quite a few tasks, it is not always the most efficient way of achieving a particular outcome, especially if a lot of calculations have to be performed to obtain a single result, while much shorter routes to this destination could be followed. On the other hand, it is even particularly challenging to achieve intelligence this way, because numbering systems are only a part of the learned, intellectual capabilities of a human. So, achieving something closer to human intelligence, will also require a platform that has another basis. Mind you that while there are currently so-called intelligent platforms around, these generally follow a quite lower standing definition of intelligence, which only corresponds with minor elements of human intelligence. The reason why the latter is more exemplar is because it manages quite complicated tasks and that with very low energy requirements, while also running in accordance with a rather slow "internal" clock.

The need for continuous improvements towards more energy efficient, intelligent devices, could be resolved through the use of probability based systems. While their underlying principle is also based on an abstraction, they are less likely to suffer from the introduction of grey "problems", since they are in essence based on the grey-principle, or more specifically the ranking of results, rather than there being a need for absolute precision, as in black / white models. However, for an abstraction to run efficiently, there is a need for an appropriate hardware platform, and that is not a platform based on number-crunching. Secondly, binary systems also lack the needed information density to deal with probability based abstractions. Additionally, another important lesson learned from the human brain lies in the use of feedback to improve the overall outcome. This feedback allows for correcting information to deal with e.g. noisy inputs. However, feedback in this context refers to the back and forth communication through different layers, which more closely aligns with what is known as reversibility within a computing context. The additional benefit of reversibility being that it brings the model closer to another feature of the human brain, namely that it creates an adiabatic computer, which will not generate any heat and consume a minimal amount of power.

While a lot of the above has not really been researched in much depth, there is a vast set of other computing research that has been performed over the recent years and the obvious question is why probability based computing would "overtake" all the other ones. Well, on performing a closer investigation, it seems that while probability based computing is not explicitly mentioned in many of the existing research, most of them could actually support probably based computing with a good information density and possibly even elements of reversibility and so on. The new paradigm, of working with imperfections, will therefore not restrict itself to the silicon era, but can certainly form a cross platform solution, which many have tried to achieve with the binary model, but did not succeed. Nevertheless, the move from a model of absolute precision towards a model of imperfect "ranking" will require a different way of development.

Developing such a new platform brings along a number of challenges, of which not all need to be dealt with together, but certainly a few are key in terms of giving the platform some of the key advantages. These key features, are: feedback (reversibility), high information density, designing with / for imperfections and timing. Each of them playing a key role of some kind, and while they can to some extend be addressed independently one should not loose sight of the overall picture, since this is part of the reason why the current platforms experience problems. In terms of using such a platform, there will also be some changes, and then more specifically to what is currently known as "programming", which will then be replaced by a learning process. Again, this brings along a number of challenges that will need to be resolved, but fortunately there are some good examples around on how we, humans, learn. After all, if the purpose is to build a system very similar to ourselves, then benefits can be drawn from decades of human evolution to help the development of these systems . One could then say that there is the risk that human-kind will have evolved much further by the time that a first platform is constructed, but at least "the human empire" will then not be threatened by "machine". Not to mention that the new programming model in combination with the different underlying platform may finally make an end to software "bugs". Unless during the steps taken to develop this new platform one would conclude that the best name for the basic cell is a bug, and then it may get famous for being the "bugs-are-everywhere"-platform.

Bibliography

[1] G. Buzsaki, *Rhythms of the Brain*. New York: Oxford Press, 2006.

[2] L. Chua, "Memristor - The Missing Circuit Element," *IEEE Transactions on Circuit Theory*, vol. 18, no. 5, pp. 507–519, 1971.

[3] R. Courtland, "The Status of Moore's Law: It's Complicated.", Available at: http://spectrum.ieee.org/semiconductors/devices/the-status-of-moores-law-its-complicated, [Accessed on: Aug 2014], 2013.

[4] D. G. Elliott, M. Stumm, W. M. Snelgrove, C. Cojocaru, and R. McKenzie, "Computational RAM: Implementing Processors In Memory," *IEEE Design & Test of Computers*, vol. 16, no. 1, pp. 32–41, 1999.

[5] M. Gokhale, B. Holmes, and K. Iobst, "Processing In Memory: the Terasys Massively Parallel PIM Array," *Computer*, vol. 28, no. 4, pp. 23–31, 1995.

[6] J. Hawkins and S. Blakeslee, *On Intelligence*. Holt Paperbacks, 2004.

[7] J. Hawkins and D. George, "Hierarchical Temporal Memory: Concepts, Theory, and Terminology," p. 20, 3/27/2007 2007.

[8] R. C. Johnson, "Missing Link, Memristor Created", Available at: http://www.eetimes.com/document.asp?doc_id=1168454, [Accessed on: Aug 2014], 2008.

[9] R. Kowalski, *Computational Logic and Human Thinking*. Cambridge University Press, 2011.

[10] J. Liu, F. Chow, T. Kong, and R. Roy, "Variable Instruction Set Architecture and its Compiler Support," *IEEE Transactions on Computers*, vol. 52, no. 7, pp. 881–895, 2003.

[11] W. J. C. Melis, P. Y. K. Cheung, and W. Luk, "Autonomous Memory Block for Reconfigurable Computing," in *Proceedings of the 2004 International Symposium on Circuits and Systems, 2004. ISCAS '04.*, vol. II, pp. 581–584.

[12] W. J. C. Melis, P. Y. K. Cheung, and W. Luk, "Scalable Structured Data Access by Combining Autonomous Memory Blocks," in *Proceedings International Conference on Field-Programmable Technology (FPT'04)*, pp. 457–460.

[13] W. J. C. Melis, S. Chizuwa, and M. Kameyama, "Evaluation of hierarchical temporal memory for a real world application," in *Proceedings of Fourth International Conference on Innovative Computing, Information and Control*, pp. 144–147.

[14] W. J. C. Melis, K. Turkington, A. Whitton, W. Luk, P. Y. K. Cheung, and P. Metzgen, "Cell Based Motion Estimators for Reconfigurable Platforms," in *Proceedings of the 2005 International Conference on Engineering of Reconfigurable Systems and Algorithms (ERSA 2005)*, pp. 218–224.

[15] P. A. Merolla, J. V. Arthur, R. Alvarez-Icaza, A. S. Cassidy, J. Sawada, F. Akopyan, B. L. Jackson, N. Imam, C. Guo, Y. Nakamura, B. Brezzo, I. Vo, S. K. Esser, R. Appuswamy, B. Taba, A. Amir, M. D. Flickner, W. P. Risk, R. Manohar, and D. S. Modha, "A Million Spiking-Neuron Integrated Circuit with a Scalable Communication Network and Interface," *Science*, vol. 345, no. 6197, pp. 668–673, 2014.

[16] K. Nishiguchi and A. Fujiwara, "Single-Electron Counting Statistics and its Circuit Application in Nano-scale Field-Effect Transistors at Room Temperature," *Nanotechnology*, vol. 20, no. 17, 2009.

[17] G. Piccinini, "Computation in Physical Systems," E. N. Zalta, Ed. The Stanford Encyclopedia of Philosophy, 2012.

[18] S. Pironio, A. Acin, S. Massar, A. Boyer de la Giroday, D. N. Matsukevich, P. Maunz, S. Olmschenk, D. Hayes, L. Luo, T. A. Manning, and C. Monroe, "Random Numbers Certified by Bell's Theorem," *Nature - Letters*, vol. 464, pp. 1021–1024, 2010.

[19] B. Vigoda, "Low Power Logic for Statistical Inference," pp. 349–354, 18-20 August 2010 2010.

[20] J. Von Neumann, *Probabilistic Logics and the Synthesis of Reliable Organisms form Unreliable Components*. Princeton University Press, 1956, pp. 43–98.

[21] J. Von Neumann, *The Computer and the Brain*, 2000.

[22] J. Webb, "Brain-Inspired Chip fits 1m Neurons on Postage Stamp," *BBC News*, Available at: http://www.bbc.co.uk/news/science-environment-28688781, [Accessed on: Aug 2014], 2014.

[23] S. Wolfram, *A New Kind of Science*, 1st ed. Wolfram Media, Inc., 2002.

I want morebooks!

Buy your books fast and straightforward online - at one of the world's fastest growing online book stores! Environmentally sound due to Print-on-Demand technologies.

Buy your books online at
www.get-morebooks.com

Kaufen Sie Ihre Bücher schnell und unkompliziert online – auf einer der am schnellsten wachsenden Buchhandelsplattformen weltweit! Dank Print-On-Demand umwelt- und ressourcenschonend produziert.

Bücher schneller online kaufen
www.morebooks.de

OmniScriptum Marketing DEU GmbH
Heinrich-Böcking-Str. 6-8
D - 66121 Saarbrücken
Telefax: +49 681 93 81 567-9

info@omniscriptum.com
www.omniscriptum.com

www.ingramcontent.com/pod-product-compliance
Lightning Source LLC
Chambersburg PA
CBHW020435220526
45464CB00002B/717